Nanostructure Science and Technology

Series Editor

David J. Lockwood, FRSC, National Research Council of Canada, Ottawa, ON, Canada

Nanostructure science and technology now forms a common thread that runs through all physical and materials sciences and is emerging in industrial applications as nanotechnology. The breadth of the subject material is demonstrated by the fact that it covers and intertwines many of the traditional areas of physics, chemistry, biology, and medicine. Within each main topic in this field there can be many subfields. For example, the electrical properties of nanostructured materials is a topic that can cover electron transport in semiconductor quantum dots, self-assembled molecular nanostructures, carbon nanotubes, chemically tailored hybrid magnetic-semiconductor nanostructures, colloidal quantum dots, nanostructured superconductors, nanocrystalline electronic junctions, etc. Obviously, no one book can cope with such a diversity of subject matter. The nanostructured material system is, however, of increasing significance in our technology-dominated economy and this suggests the need for a series of books to cover recent developments.

The scope of the series is designed to cover as much of the subject matter as possible – from physics and chemistry to biology and medicine, and from basic science to applications. At present, the most significant subject areas are concentrated in basic science and mainly within physics and chemistry, but as time goes by more importance will inevitably be given to subjects in applied science and will also include biology and medicine. The series will naturally accommodate this flow of developments in the sciences and technology of nanostructures and maintain its topicality by virtue of its broad emphasis. It is important that emerging areas in the biological and medical sciences, for example, not be ignored as, despite their diversity, developments in this field are often interlinked. The series will maintain the required cohesiveness from a judicious mix of edited volumes and monographs that while covering subfields in depth will also contain more general and interdisciplinary texts.

Thus the series is planned to cover in a coherent fashion the developments in basic research from the distinct viewpoints of physics, chemistry, biology, and materials science and also the engineering technologies emerging from this research. Each volume will also reflect this flow from science to technology. As time goes by, the earlier series volumes will then serve as reference texts to subsequent volumes.

Yanan Chen

High Temperature Shock Technology

Ultra-fast Micro-Nano Manufacturing

Springer

Yanan Chen
School of Materials Science
and Engineering
Tianjin University
Tianjin, China

ISSN 1571-5744 ISSN 2197-7976 (electronic)
Nanostructure Science and Technology
ISBN 978-981-19-8123-4 ISBN 978-981-19-8124-1 (eBook)
https://doi.org/10.1007/978-981-19-8124-1

This Springer imprint is published by the registered company Springer Nature Singapore Pte Ltd.
The registered company address is: 152 Beach Road, #21-01/04 Gateway East, Singapore 189721, Singapore

Contents

Chapter 1
Brief Introduction of HTS

Contents

1.1 Overview

In recent years, functionalized nanomaterials have drawn more and more attention to the researchers in the fields of physics, chemistry, environment, biology, medicine, etc. [1–10]. The preparation of functionalized nanomaterials can be classified into two categories: One is "from top to bottom" and the other is "from bottom to top". The method of "from top to bottom" usually includes solid-state and physical methods, for example, ball milling [11, 12], ultrasonic stripping [13, 14], electron beam irradiation [15, 16], selective surface functionalization [17], pulsed laser ablation deposition [18, 19], exploding filament [20, 21], etc. While the "from bottom to top" method mainly contains wet chemical methods (e.g., solvothermal reaction [22]) and methods of chemical treatment (e.g., spray pyrolysis [23] and chemical vapor deposition [24]). But there are still many challenges to synthesizing functional nanomaterials using the above methods, such as easy surface oxidation, easy agglomeration of particles, and low production efficiency, which hinder the commercialization of nanomaterials to a certain extent [25–27]. Moreover, traditional synthesis methods have difficulties in breaking through the inherent properties of materials. For example, traditional synthesis methods usually result in immiscible mixtures or heterogeneous structures due to thermodynamic limitations (e.g., Rh-Au bimetallic phase diagram) [28, 29].

Y. Chen, *High Temperature Shock Technology*, Nanostructure Science and Technology,
https://doi.org/10.1007/978-981-19-8124-1_1

In addition, it remains a great challenge in using traditional methods to effectively eliminate impurities and inherent defects of carbon nanomaterials (such as nanofibers and graphene) to achieve high graphitization and electrical conductivity [30, 31]. Therefore, it is tremendously urgent to develop a rapid, efficient, green, and low-cost nanomaterials preparation technology.

High Temperature Shock (HT-Shock, HTS) is a new method for ultra-fast synthesis of nanomaterials invented by Professors Yanan Chen and Liangbing Hu in 2016. Since the invention of this method, Professor Chen's team has successfully synthesized a series of high-performance micro/nanomaterials, including metal and semiconductor nanoparticles (such as aluminum, gold, platinum, silicon, palladium, etc.), alloys, high-entropy alloys, metal compounds, high-entropy compounds, graphene, ceramics, and metastable materials. At present, ultra-fast nanomaterials of HTS have shown excellent performance in energy storage and conversion, biological cryo-electron microscopy, and other application fields. More than 80 top research papers have been published in the journals of *Nature Energy, Nature Communications, Advanced Materials*, and so on, which is leading the research upsurge in the direction of ultra-fast synthesis of new materials in the world. Meanwhile, international colleagues have prepared a series of new materials utilizing this HTS technology, and their research results have been published in *Nature, Science, Nature Nanotechnology, Nature Catalyst*, and other internationally renowned journals. This book will systematically introduce the principle, development history, research status, scientific issues, application scenarios, and other aspects of ultra-fast preparation technology of micro-nanoenergy materials.

1.2 Working Principle and Device of HTS

The rapid, efficient, and new HTS technology is based on the principle of electric joule heating. The entire process of preparation lasts only a few seconds or milliseconds, the rate of heating/cooling can reach up to10^5 K/s, and the reaction temperature can reach more than 3000 K [32]. This kind of technology has been used in a variety of functional nanomaterials preparation widely in recent years, particularly in new devices, energy conversion and storage, cryo-electron microscopy, and many other fields.

The device of HTS is shown in Fig. 1.1 [32]. The whole equipment mainly consists of four parts: (1) direct current (DC) power supply, (2) heated reaction chamber, (3) spectrometer, and (4) vacuum pump. Specifically, the sample is connected to the copper conductive adhesive through silver paste and fixed on the glass bracket, while the copper conductive adhesive at both ends of the sample is connected to the DC power supply, which will cause high temperature shock when the power starts [33]. The spectrometer is used to record emission spectra and estimate the sample's temperature. By fitting the sample's emission spectrum with the blackbody radiation Eq. (1.1), the conducted sample's real-time temperature in the process of HTS can be analyzed and obtained [34, 35].

$$B_{\lambda}(\lambda, T) = \gamma \varepsilon_{\text{gray}} \frac{2hc^2}{\lambda^5} \frac{1}{e^{hc/\lambda k_B T} - 1} \quad (1.1)$$

k_B, h, c, λ, $\varepsilon_{\text{gray}}$, γ represent Boltzmann constant, Planck constant, light speed, wavelength, stable emissivity, and fitting constant, respectively.

In the case of different electrical power input, the real-time temperature of sample can be accurately calculated according to the fitting between wavelength and spectral emissivity (Fig. 1.2a, b) [32]. Turn on the DC power supply and increase the electric power, the brightness of the sample is enhanced, which indicates that the sample's real-time temperature increases (Fig. 1.2c) [36]. Turn off the power, the joule heating process is stopped and the sample will be cooled to room temperature just in a few milliseconds.

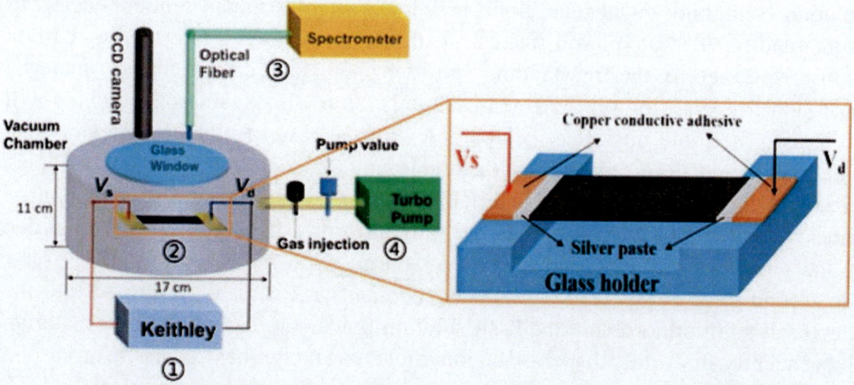

Fig. 1.1 Schematic diagram of the device of HTS

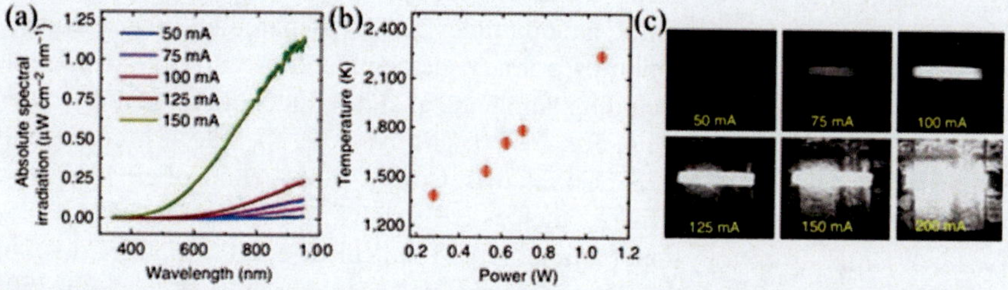

Fig. 1.2 **a** Luminescence spectra of synthetic materials at different current densities; **b** room temperature of synthetic materials at different electric powers; **c** photographic images of high temperature shock and joule heating process at different electric powers (width: 380 mm, length: 1917 mm)

1.3 Growth Mechanism of Functional Nanomaterials

1.3.1 Preparation of Nanoparticles

In order to synthesize uniformly dispersed nanomaterials, a critical step is to combine the self-supporting substrate materials and the precursors through feasible and simple pretreatment. Pretreated conductive carbon carriers, such as carbon nanofiber (CNF) [28, 37], reduced graphene oxide (RGO) films [32, 38], carbonized biomass materials (i.e., wood) [39, 40], etc., can be used as independent substrate materials and are important parts of the heating device in the reaction chamber. As shown in Fig. 1.3a, b, precursor materials can be mainly classified into two types: (1) solid materials with micron size, such as semiconductor, metal, and compound powder; (2) salt solution containing metal ions. Solid precursors such as metal, semiconductor, or compound powders in micron size can be directly placed on a freestanding carbon matrix, and then fix the freestanding "powder precursor - carbon matrix" material on a concave glass bracket. After high temperature shock, micron particles will be transformed into ultra-fine nanoparticles. In the case of metal precursors, take "metal precursor-RGO films" as an example; micron-scale metal precursors (such as nickel, aluminum, tin, gold, and palladium) first melt during high temperature shock, and after that, they self-assemble into uniformly dispersed nanoparticles due to the separation of defects in RGO films during ultra-rapid cooling [33]. As for the second kind of precursor, the conductive carbon carrier is infiltrated into the metal salt solution to obtain the "salt solution precursors-carbon matrix" material. Subsequently, uniformly dispersed nanoparticles can be synthesized in situ on carbon matrix by HTS treatment. Specifically, the ultra high temperature generated by HTS can decompose metal salt precursors into metal clusters first, and then the metal clusters solidify into uniformly dispersed nanoparticles on the carbon carrier within a few milliseconds [41, 42] during the rapid cooling process. In addition, using ultra-fast HTS technology, monatomic nanoparticles can be prepared by precise control of the melting chemical reaction of primary elements and precursors. At present, a variety of nanomaterials, including single metal and semiconductor nanoparticles (such as Sn, Si, Ag, Au, Ir, Pd, Ru, Pt, Ni, and Al) [32, 37, 38, 43], compound nanoparticles (such as SiC, FeS_2, CoS, Co_2B, Co_3O_4, MoS_2, and $CoFeP_x$) [40, 44–47], bimetallic alloy nanoparticles (such as NiFe, PdNi, PtFe, and $Cu_{0.9}Ni_{0.1}$) [39, 42, 48, 49], mixing entropy nanoparticles (e.g., PtPdRhRuCe, PtPdCoNiFeCuAuSn, and CoMoFeNiCu) [28, 34], and monatomic materials (e.g., Co, Pt, and Ru) [50] have been successfully prepared by HTS.

1.3.2 Preparation of Carbon-Based Materials

In addition to the preparation of nanoparticles described above, HTS can be utilized to prepare carbon-based nanomaterials directly, which include carbon nanotube (CNT)

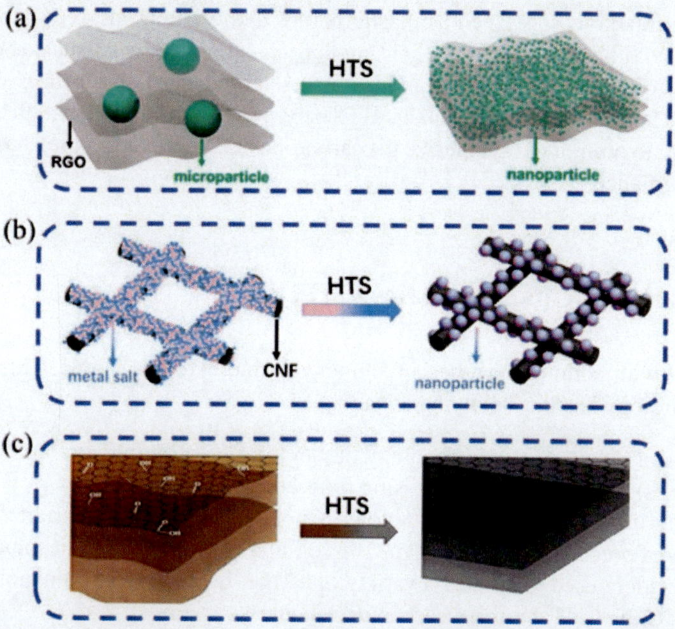

Fig. 1.3 Preparation of functional nanomaterials by HTS **a** schematic diagram of the process by which micron-sized solid materials (metals, semiconductors, and compound powders) are assembled into nanoparticles; **b** schematic diagram of the converting a salt solution containing metal ions into nanoparticles process; **c** schematic diagram of the preparation process of carbon-based nanomaterials

[30, 51], CNF [52], RGO [31, 36], carbonized switchgrass [53, 54], lignin carbon, and three-dimensional (3D) printing carbon-based materials [55], etc. (Fig. 1.3c). It is worth noticing that the physicochemical properties of carbon-based nanomaterials (such as graphitization degree, thermal, and electrical conductivity) can get a significant improvement by ultra high temperature treatment during high temperature shock. For example, RGO films treated with HTS have a more compact structure, fewer defects/impurities, and larger graphite sp^2 region, resulting in high conductivity ($\sigma = 6300$ S/cm) and carrier mobility ($\mu = 320\ \text{cm}^2/(\text{V s})$) and high thermoelectric power factor $S^2\sigma = 54.5\ \mu\text{W}/(\text{cm K}^2)$ (S stands for Seebeck coefficient) [31, 56]. Moreover, CNF/CNT can be effectively grown into a 3D continuous cross-linked network structure with strong electrical conductivity by HTS. The whole process does not require any other crosslinking agents, functional monomers, or costly facilities. HTS with the characteristics of low-cost and ultra-fast preparation can achieve covalent crosslinking between adjacent CNTs in milliseconds. Compared with polyacrylonitrile-based CNT, CNT treated by HTS had specific conductivity (1060 (S cm^2)/g) and higher specific strength (2.2 N/tex) [51]. Moreover, lignin-based biomass treated by HTS can be effectively converted into highly crystalline graphite carbon. This kind of technology can turn "waste" into "treasure" easily by

converting mixed wastes (such as plastic bottles and waste food), recyclable products (such as rubber tires and biochar), and cheap carbon sources (such as coal and petroleum coke) into high-quality and large-scale graphene materials in a very short time [57]. Therefore, it can be seen that HTS can be used as a method of high temperature heating to promote to synthesize the carbon-based functional nanomaterials with high performance.

1.4 Technical Characteristics of HTS

Compared with common synthesis methods of nanomaterials (Table 1.1), the new HTS technology has the following advantages:

(1) Fast and efficient. This technology has an ultra-fast heating/cooling rate (up to 10^5 K/s) and ultra-short processing time (milliseconds);
(2) Strong universality. This technology is not only widely appropriate for many kinds of precursors (including micron solid powder and metal salt solution), but also widely used to prepare a variety of carbon-based functional nanomaterials (e.g., CNF, CNT, biomass carbon, RGO, etc.);
(3) This technology can effectively avoid surface oxidation, immiscibility, and agglomeration in the nanomaterials preparation process and can eliminate the defects or impurities in carbon-based nanomaterials;
(4) Functional nanomaterials which are prepared by HTS show excellent physical and chemical properties (e.g., ultra-fine particle size, high activity, uniform distribution, fewer defects, or impurities) and thus have excellent electrochemical properties;

Table 1.1 Comparison of HTS with other common nanomaterials synthesis methods

	Solvothermal	Ball mill	Tube furnace annealing	Spray pyrolysis	Laser ablation deposition	HTS
Time scale	h	h	h	s	ms	ms
Heating rate (K/s)	<10	–	<10	<10	>10^6	up to 10^5
Cost	Low	High	Mid	Mid	High	Low
Environmental friendliness	Mid	High	Mid	Mid	High	High
Energy consumption	Mid	High	High	High	High	Low
Particle dispersion	Low	Low	Low	Mid	Low	High
Particle mixability	Low	Low	Low	Low	Low	High

(5) Industrializable. For example, combining HTS technology with roll-to-roll synthesis methods could enable the large-scale production of nanomaterials in the field of energy conversion and storage in industry.

In recent years, a series of studies using HTS technology in energy applications have been reported, and it still has great potential which can be exploited in the future. Rapid and efficient HTS is a unique and original method to prepare a variety of nanomaterials for energy conversion and storage. Up to now, a variety of carbon-based nanomaterials with high quality and functional nanoparticles have been successfully synthesized through this technology and are applied in smart devices, batteries, and electro-catalysis. Due to the universality of HTS technology, this method not only can prepare and synthesize various nanoparticles (containing single or mixed metal alloy, metal sulfides, metal oxides, metal boride, metal selenide, metal phosphide, etc.), but also can be used for the synthesis of RGO, CNF, CNT, switchgrass, carbon cloth, biomass carbon materials, and carbon nanomaterial. Furthermore, this technology can be extended to the preparation of a range of bulk or materials which are sensitive to temperature (e.g., glass, ceramics, plastics, steel, organic, metal–organic skeleton, and polymer materials). Amorphous nanomaterials have attracted a lot of attention because of their unique properties such as defect distribution, loose structure, and isotropy. Many studies have shown that amorphous nanomaterials show enhanced electrochemical properties such as rate performance and long cycle stability of lithium/sodium batteries [58–62]. Besides, amorphous catalysts have more catalytic active sites than highly crystalline nanomaterials due to the increased number of defects [63–65]. By optimizing HTS processes (e.g., using water or liquid nitrogen to regulate heating or cooling rates), a range of high-performance amorphous/amorphous heterojunction nanomaterials can be designed and synthesized. Because of this, it is believed that HTS technology will attract more and more attention in the future. In application, HTS technology is helpful to the development of batteries and electro-catalysis. It shows great application potential in different battery systems (such as lithium/sodium/potassium ion battery, lithium/sodium/potassium metal battery, supercapacitor, lithium/sodium/potassium ion hybrid capacitor, lithium/sodium/potassium–sulfur battery) and catalytic system (such as CO oxidation, CO_2 reduction, water decomposition, fuel cell). In addition, this technology offers great possibilities for large-scale intelligent applications such as ultra high temperature heating, thermoelectric power conversion, thermal management, temperature sensors, wearable lighting, flexible electronics, and mechanical switches. At present, artificial intelligence has become an effective method to speed up the high-performance new materials development [66], and rapid and efficient HTS technology can achieve high-throughput synthesis of various nanomaterials. As a result, combining HTS technology with artificial intelligence to synthesize and identify the best performance of nanomaterials is a promising application prospect. In addition to the above-mentioned aspects, what is worth mentioning is that the heater in the reaction chamber of HTS synthesis device is an indispensable and important component. In addition to the most widely used freestanding conductive carbon substrate as heater, a couple of heaters have recently been developed for high temperature thermal impact heating units. For example, quartz tube cylindrical heaters between two copper electrodes, and two highly flexible carbon strip sandwich heaters. In order to realize the rapid preparation of different functional materials, innovative self-made heaters with various characteristics should be designed.

At the same time, HTS technology still faces some challenges, which are mainly reflected in the following parts:

(1) The major challenge in the synthesis of nanomaterials by HTS lies in the accurate control of their structure and morphology. The nanomaterials prepared by this technology generally show the shape of solid sphere with nanometer size, but lack the complex layered structure (e.g., core–shell, hollow, yolk shell, and multi-shell structure) which limits the further improvement of their performance in the field of energy storage. Here, the rational design of HTS processes (e.g., changes in the external environment, uses secondary treatment, and additives) is thought to induce to synthesize the functional nanomaterials with well-defined sizes, shapes, structures, and crystal phases;
(2) There are still challenges in enhancing the mass load of nanomaterials. We know that high-quality loaded active nanostructured electrodes are important for practical applications in the devices of energy conversion and storage. Though the nanomaterials synthesized by HTS have self-supporting structure, enhancing the mass load might still be difficult. Therefore, the mass load of functional nanomaterials can be further improved by fine-tuning the precursors and parameters of the HTS process and designing a homemade heater that can directly process the precursors without the need for an independent conductive carbon matrix;
(3) Study the behavior and mechanism of HTS is a great challenge. It is very difficult to observe the formation of nanomaterials because of the extremely short processing time. However, the development of in situ characterization technology and theoretical research will be conducive to understanding the formation mechanism.

In conclusion, the new HTS technology with rapid, efficient, green, and low cost has great application potential, especially plays an important role in the rational design and preparation of multi-functional nanomaterials. It can be predicted that HTS technology has a wide range of applications and prospects in energy conversion and storage, carbon materials, and new devices and will provide new chances for developing the next generation of energy-related equipment.

References

1. Guo Y, Xu K, Wu C et al (2015) Surface chemical-modification for engineering the intrinsic physical properties of inorganic two-dimensional nanomaterials. Chem Soc Rev 44(3):637–646
2. Shin TH, Cheon J (2017) Synergism of nanomaterials with physical stimuli for biology and medicine. Acc Chem Res 50(3):567–572
3. Ray PC (2010) Size and shape dependent second order nonlinear optical properties of nanomaterials and their application in biological and chemical sensing. Chem Rev 110(9):5332–5365

4. Sun Y, Liu N, Cui Y (2016) Promises and challenges of nanomaterials for lithium-based rechargeable batteries. Nat Energy 1:1–12
5. Mei J, Liao T, Kou L et al (2017) Two-dimensional metal oxide nanomaterials for next-generation rechargeable batteries. Adv Mater 29(48):1700176
6. Cong L, Xie H, Li J (2017) Hierarchical structures based on two-dimensional nanomaterials for rechargeable lithium batteries. Adv Energy Mater 7(12):1601906
7. Perreault F, De Faria AF, Elimelech M (2015) Environmental applications of graphene-based nanomaterials. Chem Soc Rev 44(16):5861–5896
8. Prieto G, Zečević J, Friedrich H et al (2013) Towards stable catalysts by controlling collective properties of supported metal nanoparticles. Nat Mater 12(1):34–39
9. Liu Y, Dong X, Chen P (2012) Biological and chemical sensors based on graphene materials. Chem Soc Rev 41(6):2283–2307
10. Lee JH, Huh YM, Jun YW et al (2007) Artificially engineered magnetic nanoparticles for ultra-sensitive molecular imaging. Nat Med 13(1):95–99
11. Zhang H, Yu J, Chen Y et al (2006) Conical boron nitride nanorods synthesized via the ball-milling and annealing method. J Am Ceram Soc 89(2):675–679
12. Jeon IY, Bae SY, Seo JM et al (2015) Scalable production of edge-functionalized graphene nanoplatelets via mechanochemical ball-milling. Adv Func Mater 25(45):6961–6975
13. Yadav V, Roy S, Singh P et al (2019) 2D MoS_2-based nanomaterials for therapeutic, bioimaging, and biosensing applications. Small 15(1):1803706
14. Ling S, Li C, Jin K et al (2016) Liquid exfoliated natural silk nanofibrils: applications in optical and electrical devices. Adv Mater 28(35):7783–7790
15. Kim JU, Cha SH, Shin K et al (2005) Synthesis of gold nanoparticles from gold (I)−alkanethiolate complexes with supramolecular structures through electron beam irradiation in TEM. J Am Chem Soc 127(28):9962–9963
16. Gao Y, Bando Y (2002) Carbon nanothermometer containing gallium. Nature 415(6872):599–599
17. Park I, Li Z, Pisano AP et al (2007) Selective surface functionalization of silicon nanowires via nanoscale joule heating. Nano Lett 7(10):3106–3111
18. Pronko P, Zhang Z, VanRompay P (2003) Critical density effects in femtosecond ablation plasmas and consequences for high intensity pulsed laser deposition. Appl Surf Sci 208:492–501
19. Orii T, Hirasawa M, Seto T (2003) Tunable, narrow-band light emission from size-selected Si nanoparticles produced by pulsed-laser ablation. Appl Phys Lett 83(16):3395–3397
20. Sahai A, Goswami N, Kaushik S et al (2016) $Cu/Cu_2O/CuO$ nanoparticles: novel synthesis by exploding wire technique and extensive characterization. Appl Surf Sci 390:974–983
21. Sindhu T, Sarathi R, Chakravarthy SR (2007) Understanding nanoparticle formation by a wire explosion process through experimental and modelling studies. Nanotechnology 19(2):025703
22. Murukanahally D, Itaru H (2012) Hydrothermal and solvothermal process towards development of $LiMPO_4$ (M=Fe, Mn) nanomaterials for lithium-ion batteries. Adv Energy Mater 2(3):284–297
23. Jin L, Zhixing W, Jiexi W et al (2019) Advances in nanostructures fabricated via spray pyrolysis and their applications in energy storage and conversion. Chem Soc Rev 48(11):3015–3072
24. Zhang Y, Zhang L, Zhou C (2013) Review of chemical vapor deposition of graphene and related applications. Acc Chem Res 46(10):2329–2339
25. Maillard F, Schreier S, Hanzlik M et al (2005) Stimming, Influence of particle agglomeration on the catalytic activity of carbon-supported Pt nanoparticles in CO monolayer oxidation. Phys Chem Chem Phys 7(2):385–393
26. Chen X, Wu G, Chen J et al (2011) Synthesis of "clean" and well-dispersive Pd nanoparticles with excellent electrocatalytic property on graphene oxide. J Am Chem Soc 133(11):3693–3695
27. Ganguli AK, Ganguly A, Vaidya S (2010) Microemulsion-based synthesis of nanocrystalline materials. Chem Soc Rev 39(2):474–485
28. Yao Y, Huang Z, Xie P et al (2018) Carbothermal shock synthesis of high-entropy-alloy nanoparticles. Science 359(6383):1489–1494

29. Chen PC, Liu X, Hedrick JL et al (2016) Polyelemental nanoparticle libraries. Science 352(6293):1565–1569
30. Yao Y, Jiang F, Yang C et al (2018) Epitaxial welding of carbon nanotube networks for aqueous battery current collectors. ACS Nano 12(6):5266–5273
31. Wang Y, Chen Y, Lacey SD et al (2018) Reduced graphene oxide film with record-high conductivity and mobility. Mater Today 21(2):186–192
32. Chen Y, Egan GC, Wan J et al (2016) Ultra-fast self-assembly and stabilization of reactive nanoparticles in reduced graphene oxide films. Nat Commun 7:12332
33. Li Y, Chen Y, Nie A et al (2017) In Situ, fast, high-temperature synthesis of nickel nanoparticles in reduced graphene oxide matrix. Adv Energy Mater 7(11):1601783
34. Xie P, Yao Y, Huang Z et al (2019) Highly efficient decomposition of ammonia using high-entropy alloy catalysts. Nat Commun 10:4011
35. Bao W, Pickel AD, Zhang Q et al (2016) Flexible, high temperature, planar lighting with large scale printable nanocarbon paper. Adv Mater 28(23):4684–4691
36. Chen Y, Fu K, Zhu S et al (2016) Reduced graphene oxide films with ultrahigh conductivity as Li-ion battery current collectors. Nano Lett 16(6):3616–3623
37. Yang C, Yao Y, He S et al (2017) Ultrafine silver nanoparticles for seeded lithium deposition toward stable lithium metal anode. Adv Mater 29(38):1702714
38. Chen Y, Li Y, Wang Y et al (2016) Rapid, in situ synthesis of high capacity battery anodes through high temperature radiation-based thermal shock. Nano Lett 16(9):5553–5558
39. Li Y, Gao T, Yao Y et al (2018) In Situ "Chainmail Catalyst" assembly in low-tortuosity, hierarchical carbon frameworks for efficient and stable hydrogen generation. Adv Energy Mater 8(25):1801289
40. Yang C, Cui M, Li N et al (2019) In situ iron coating on nanocatalysts for efficient and durable oxygen evolution reaction. Nano Energy 63:103855
41. Yao Y, Chen F, Nie A et al (2017) In Situ high temperature synthesis of single-component metallic nanoparticles. ACS Cent Sci 3(4):294–301
42. Chen F, Yao Y, Nie A et al (2018) High-temperature atomic mixing toward well-dispersed bimetallic electrocatalysts. Adv Energy Mater 8(25):1800466
43. Yao Y, Huang Z, Xie P et al (2019) Ultrafast, controllable synthesis of sub-nano metallic clusters through defect engineering. ACS Appl Mater Interfaces 11(33):29773–29779
44. Xie H, Fu K, Yang C et al (2018) Necklace-like silicon carbide and carbon nanocomposites formed by steady joule heating. Small Meth 2(4):1700371
45. Chen Y, Xu S, Li Y et al (2017) FeS_2 nanoparticles embedded in reduced graphene oxide toward robust, high-performance electrocatalysts. Adv Energy Mater 7(19):1700482
46. Chen Y, Xu S, Zhu S et al (2019) Millisecond synthesis of CoS nanoparticles for highly efficient overall water splitting. Nano Res 12(9):2259–2267
47. Xu S, Chen Y, Li Y et al (2017) Universal, in situ transformation of bulky compounds into nanoscale catalysts by high-temperature pulse. Nano Lett 17(9):5817–5822
48. Yang C, Ko B H, Hwang S et al (2020) Overcoming immiscibility toward bimetallic catalyst library. Sci Adv 6(17):eaaz6844
49. Zeng S, Lv B, Qiao J et al (2019) PtFe alloy nanoparticles confined on carbon nanotube networks as air cathodes for flexible and wearable energy devices. ACS Appl Nano Mat 2(12):7870–7879
50. Yao Y, Huang Z, Xie P et al (2019) High temperature shock wave stabilized single atoms. Nat Nanotechnol 14(9):851–857
51. Song Y, Di J, Zhang C et al (2019) Millisecond tension-annealing for enhancing carbon nanotube fibers. Nanoscale 11(29):13909–13916
52. Yao Y, Fu KK, Zhu S et al (2016) Carbon welding by ultrafast joule heating. Nano Lett 16(11):7282–7289
53. Zhang F, Yao Y, Wan J et al (2017) High temperature carbonized grass as a high performance sodium ion battery anode. ACS Appl Mat Inter 9(1):391–397
54. Jiang F, Yao Y, Natarajan B et al (2019) Ultrahigh-temperature conversion of biomass to highly conductive graphitic. Carbon 144:241–248

55. Yao Y, Fu KK, Yan C et al (2016) Three-dimensional printable high-temperature and high-rate heaters. ACS Nano 10(5):5272–5279
56. Li T, Pickel AD, Yao Y et al (2018) Thermoelectric properties and performance of flexible reduced graphene oxide films up to 3000 K. Nat Energy 3(2):148–156
57. Luong DX, Bets KV, Algozeeb WA et al (2020) Gram-scale bottom-up flash graphene synthesis. Nature 577(7792):647–651
58. Li J, Doubek G, McMillon-Brown L et al (2019) Recent advances in metallic glass nanostructures: synthesis strategies and electrocatalytic applications. Adv Mater 31(7):1802120
59. Yan S, Abhilash K, Tang L et al (2019) Research advances of amorphous metal oxides in electrochemical energy storage and conversion. Small 15(4):1804371
60. Wonterghem J, Mørup S, Koch CJ et al (1986) Formation of ultra-fine amorphous alloy particles by reduction in aqueous solution. Nature 322(6080):622–623
61. Fan L, Li X, Yan B et al (2016) Controlled SnO_2 crystallinity effectively dominating sodium storage performance. Adv Energy Mater 6(10):1502057
62. Idota Y, Kubota T, Matsufuji A et al (1997) Tin-based amorphous oxide: a high-capacity lithium-ion-storage material. Science 276(5317):1395–1397
63. Wu G, Zheng X, Cui P et al (2019) A general synthesis approach for amorphous noble metal nanosheets. Nat Commun 10:4855
64. Morales-Guio CG, Hu X (2014) Amorphous molybdenum sulfides as hydrogen evolution catalysts. Acc Chem Res 47(8):2671–2681
65. He D, Zhang L, He D et al (2016) Amorphous nickel boride membrane on a platinum–nickel alloy surface for enhanced oxygen reduction reaction. Nat Commun 7:12362
66. Butler KT, Davies DW, Cartwright H et al (2018) Machine learning for molecular and materials science. Nature 559(7715):547–555

Chapter 2
Application of HTS in Green Hydrogen and Fuel Cells

Contents

2.1 Introduction

The methods of catalysts synthesis reported in the past, such as electrodeposition, hydrothermal, coprecipitation, etc., were mostly limited to the preparation of materials with thermodynamic advantages (e.g., heterogeneous structure), which hindered the development of catalytic materials [1–4]. In order to break through the restrictions of the methods of conventional synthesis, the necessity is to exploit a range of ultra-fast synthesis strategies which are able to adapt to the tough conditions of the kinetic. Related strategies have been widely concerned by researchers recently, such as spray pyrolysis (SP), microwave (MW), laser ablation in liquid (LAL), and high temperature shock (HTS) [5–9].

Nanomaterials with a variety of morphology, size, and composition are able to offer rich sites to perform surface chemical reactions, which shows huge practical potential in industry. A few metastable nanomaterials were synthesized under

Y. Chen, *High Temperature Shock Technology*, Nanostructure Science and Technology,
https://doi.org/10.1007/978-981-19-8124-1_2

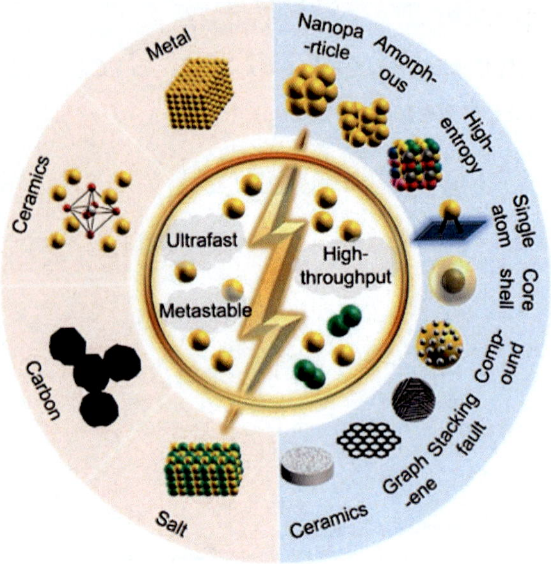

Fig. 2.1 Precursors and various products schemes of HTS

extreme circumstances, and they have special chemical and physical characteristics that traditional materials do not possess. Metastable nanoparticles which have abundant defects are widely considered as candidates with great promise for highly active catalysts. As aforementioned, ultra-fast synthesis strategies can provide a new approach to break though the restrictions of traditional methods in producing metastable nanomaterials. HTS is a method to obtain nanomaterials by heating the sample to super high temperature in a very short time and then cooling it rapidly. As a new manufacturing method, it is expected to convert various salt precursors or bulk materials into nanomaterials instantly, including highly oriented materials, core–shell structure, nanoparticles, nanoparticles rich in lamination, single atoms, and compounds, as shown in Fig. 2.1. This is due to the characteristics of HTS method: (1) The heating speed can reach up to 10^5 K/s. Extremely fast heating speed can make the system reach ultra high temperature in a short time; (2) short heating time can prevent further growth and coarsening of the alloy particles at high temperature; (3) the ultra-fast cooling rate of strong quenching provides sufficient dynamic conditions for metastable alloy particles preparation, for example, the cooling rate can reach 10^{14} K/s.

2.2 Electrocatalytic Reaction

Although many chemical reactions are thermodynamically feasible, they themselves do not occur at significant macroscopic rates and therefore require the use of catalysts to lower the energy of reaction activation and increase the reaction speed. Electrocatalytic reaction is based on electrochemical reaction. The catalytic material is used as electrode or the catalyst is loaded on the electrode surface, so as to reduce the activation energy of reaction and improve the efficiency of electrochemical reaction. Electrocatalysis is characterized by the reaction process consisting of more than two successive steps and the formation of chemisorbed intermediates on the electrode surface. Many of the most important electrode reactions that generate or degrade molecules from ions are electrocatalytic reactions, which fall into two main categories: The first type is that ions or molecules generate chemisorption intermediates on the electrode surface through electron transfer steps, and then the intermediates generate stable molecules through heterogeneous chemical steps or electrochemical desorption steps; the second type of reaction is the dissociative or associative chemisorption of the reactants on the electrode surface, followed by electron transfer or surface chemical reaction of the intermediates or adsorbed reactants. The first type of reaction includes hydrogen electrode process [e.g., hydrogen evolution reaction (HER) and hydrogen oxidation reaction (HOR)] and oxygen electrode process [e.g., oxygen evolution reaction (OER) and oxygen reduction reaction (ORR)], etc. The reaction of hydrogen electrode is very important and can be used as: (1) To construct reference electrode, for example, standard hydrogen electrode (SHE) and reversible hydrogen electrode (RHE); (2) hydrogen desorption reaction is very important for the development of electrochemical theory; (3) electrolysis, electroplating, and other important electrochemical processes include hydrogen precipitation reaction; (4) hydrogen anodic oxidation is the anodic reaction of proton-exchange membrane fuel cell. For oxygen electrode process, ORR is the cathode reduction reaction of fuel cell. Kinetics and mechanism have always been an important research topic in electrochemical field. The second type of reaction includes organic small molecule oxidation reaction, such as methanol, formic acid, ethylene glycol, etc. The design and preparation of catalyst has become one of the important ways to improve the performance of organic small molecule direct fuel cell. Catalysts are the core of electrocatalysis research. The important task of electrocatalysis research is to design and prepare electrocatalysts with high activity, high stability, high selectivity, and long life for specific reactions. Under the condition of electrocatalysis, except by controlling the electrode potential to control oxidation or reduction, the electric catalyst surface structure (chemical structure, atomic arrangement structure, and electronic structure) and interface structure of electric double layer also have a direct effect on the electrical efficiency and selectivity of catalytic reaction. Regulation of the interaction between electrocatalyst and reaction molecules can reduce the activation energy or change the reaction pathway. Therefore, methods just like defect generation, cation and anion regulation, strain engineering, and heteroatomic doping are suitable to adjust the catalysts' electronic structure and produce lots of active sites, thus improving the catalytic performance.

2.3 Preparation of Electrolytic Water Catalyst by HTS

Increasing energy demand and environmental pollution have prompted researchers to develop a carbon–neutral energy economy. The generation of sustainable and environmentally friendly energy is the lifeline of modern civilization. Hydrogen is produced commercially by steam reforming technology, namely gasification of coal and other heavy hydrocarbons and steam reforming of natural gas or other light hydrocarbons. This method of hydrogen production not only emits greenhouse gases, but also consumes a large amount of energy, which also has a negative impact on the environment [10]. It has been proved that hydrogen by water electrolysis is a promising energy conversion technology which can achieve environmentally friendly energy. Electrocatalytic water decomposition to produce molecular hydrogen using cheap, non-precious metals, and environmentally friendly materials is a promising and sustainable method that minimizes dependence on fossil fuels and society's carbon footprint [11–13].

In the hydrolysis system, HER occurs at the cathode to produce hydrogen, and OER occurs at the anode to produce oxygen. An external current is used to overcome the energy barrier of the reaction [14].

Under acidic conditions:

$$\text{Anode}: 2H_2O \rightarrow 4H^+ + O_2 + 4e^- \, E_a^0 = 1.23\ \text{V}$$

$$\text{Cathode}: 4H^+ + 4e^- \rightarrow 2H_2 \, E_c^0 = 0\ \text{V}$$

Under alkaline conditions:

$$\text{Anode}: 4OH^- \rightarrow 2H_2O + O_2 + 4e^- \, E_a^0 = -0.40\ \text{V}$$

$$\text{Cathode}: 4H_2O + 4e^- \rightarrow 2H_2 + 4OH^- \, E_c^0 = -0.83\ \text{V}$$

Whether in an acidic environment or an alkaline environment, the total reaction equation for electrolysis of water is: $2H_2O \rightarrow H_2 + O_2$.

Although the principle of water electrolysis is straightforward, the production of hydrogen by water electrolysis is still not widely used. The earliest commercial applications of hydroelectricity date back to the 1890s, and after more than 100 years, using the method of water electrolysis to produce hydrogen accounts for only 4% of the hydrogen supply in the world. In practical application, the efficiency of the energy conversion of electrolytic system is approximately 56–73%, which limits a wide range of application greatly [15, 16]. This is because the HER at the cathode is a reaction of two-electron transfer, while the OER at the anode is a reaction of four-electron proton coupling. The mechanism is complex, the reaction kinetics is slow, and higher energy is required. The theoretical decomposition voltage of water is 1.23 V (25 °C, 1 atm). In order to decompose water at the actual rate, an additional voltage must be applied to the cell. The total operating voltage (V_{op}) of water decomposition can be described as: $V_{op} = 1.23\ \text{V} + |\eta_c| + \eta_a + \eta_\Omega$.

η_Ω represents the excess potential used to make up for the system internal resistance (e.g., contact resistance and solution resistance); η_c and η_a are the overpotentials which are requested to overcome the cathode/anode intrinsic activation barrier. Therefore, the design and synthesis of efficient OER catalyst is the key to decrease the intrinsic activation potential barrier, improve the efficiency, and promote the wide application of electrolytic water hydrogen production. Platinum (Pt) was found to be the most effective HER electrocatalyst, while oxides of precious metals iridium (Ir) and ruthenium (Ru) (IrO_2, RuO_2, etc.) were considered to be the best OER catalysts. At present, transition metal (Ni, Fe, Co, Mn, etc.) catalysts are also a very attractive research direction.

2.3.1 Preparation of HER Catalyst by HTS

With the continuous growth of global energy demand and the increasingly serious environmental problems related to the consumption of energy, the development and production of sustainable and clean energy is urgent. Hydrogen is a promising substitute energy source with high energy density and a completely clean combustion process [17, 18]. Electrochemical/photochemical decomposition of water is one of the most appealing hydrogen production strategies currently available for hydrogen fuel production. At present, the most advanced catalyst for HER is precious metal platinum (Pt), whose overpotential is only about 50 mV when the current density reaches 100 mA/cm^2. However, the high cost and low element abundance of this precious metal have prevented its large-scale production and application as a catalyst for hydrogen production. It is of great significance to develop catalysts with the characteristic of high stability, high activity, and low cost for water electrolysis technology.

Because of the slow kinetics of HER, the improvement of the performance of hydrogen production is limited by stability and overpotential. Strain engineering is known as an effective strategy for dynamically enhancing electrochemical reactions which can optimize the chemical activity and electronic structure of catalysts [19–24]. An effective way to fabricate strained atomic structures is to fabricate catalysts with abundant defects in atomic structures, including surface vacancies, dopants, dislocations, grain boundaries, etc. [25]. Traditional methods for synthesizing strain-effecting structures include seed-mediated growth, electrochemical dealloying, thermal annealing-induced segregation, polyol synthesis, and galvanic replacement, which are low manufacturing efficient, time-consuming, and complex thus restricting the widespread application of the strain-effect catalysts [26–31]. Furthermore, strain-induced high-energy surface structures caused by defects (e.g., grain boundaries or dislocations) tend to resist surface recombination more during catalytic processes. It is generally believed that non-equilibrium conditions often lead to many defects, and the HTS method can provide an extreme environment to achieve the defect strain structure of the catalyst.

Yanan Chen/Yida Deng/Wenbin Hu's team achieved the nanofabrication of dislocation strain IrNi (DSIrNi) alloy nanoparticles through the unsteady HTS process under extreme environment [32]. In general, during conventional annealing, dislocations have sufficient time to move to the crystal's edge and then disappear, few dislocations remain in the crystal, and HTS breaks the limit on the number of dislocations. During HTS treatment, $IrCl_3$ and $NiCl_2$ are decomposed into atoms Ir and Ni, which undergo ultra-fine coagulation and crystallize in about 150 ms. The resulting IrNi nanoparticles are distributed uniformly throughout the carbon nanotube sponge (DSIrNi@CNTs). The surface electron distribution of atom Ir can be optimized, and the change of Ir atom electronic structure will have an influence on the adsorption/desorption energy of Ir atom for intermediate products of catalytic reaction, thus affecting the catalytic activity. The dislocation caused by HTS is shown in Fig. 2.2. Temperature changes over time as shown in the center of Fig. 2.2a, a series of images of infrared temperature were obtained through an infrared imaging device, indicating the heating process of DSIrNi@CNTs film as the electrical power applied increases. The sample was heated to about 1300 K, and then the input current was directly cut off for rapid quenching. The rates of heating and cooling were about 6100 K/s and 5100 K/s, respectively. The dislocation is preserved in this ultra-fast process, where the extra half plane in the dislocation (shown in yellow) forms strain fields in the IrNi nanoparticles, resulting in high-energy surfaces for catalysis (Fig. 2.2a). Figure 2.2b presents the structural modification effect on dislocation motion. Because of the process of millisecond quenching and the different solute atoms radii, dislocation motion is impeded by lattice distortion and trapped in the crystal, so a huge number of dislocations can be retained by HTS treatment.

The original carbon nanotubes (CNTs) morphology and the synthesized DSIrNi@CNTs are shown in Fig. 2.3. TEM images shown in Fig. 2.3d indicate that most nanoparticles stabilized on CNTs are subspherical or spherical and distributed uniformly. Besides, IrNi nanoparticles coated with a thin carbon layer (Fig. 2.3e) prevents IrNi nanoparticles agglomeration and improves the performance of catalytic.

The electrochemical activity evaluation of DSIrNi@CNTs was shown in Fig. 2.4. Electrochemical results showed that DSIrNi@CNTs has competitive activity for HER, achieving a current density of 10 mA/cm^2 at overpotential of 17 mV, and has excellent stability that exceeds most of the highly efficient catalysts of HER previously reported in alkaline solutions, containing the most advanced Pt/C catalysts. The preparation of DSIrNi@CNTs by HTS paves the way for designing efficient electrocatalysts.

In the last few decades, researchers have attempted to reduce the cost of electrocatalysts in two ways, one by reducing the precious metal load in electrocatalysts, and the other by developing precious metal-free electrocatalysts. Such as DSIrNi@CNTs synthesized above, the addition of nickel can not only improve the catalytic activity, but also reduce the amount of precious metal Ir, thus reducing the cost.

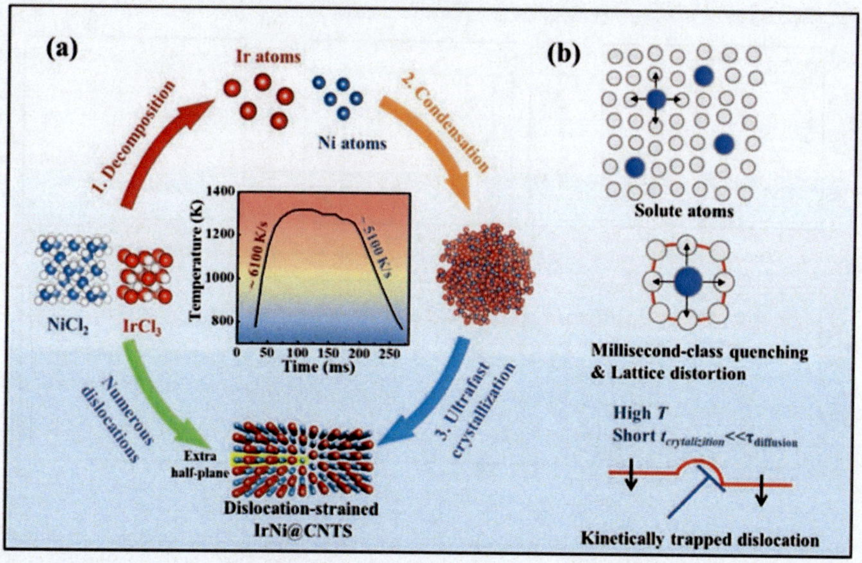

Fig. 2.2 Dislocations generated during HTS **a** $IrCl_3$ and $NiCl_2$ decompose into Ir and Ni atoms, and then undergo ultra-fine condensation and crystallization in about 150 ms. The heating and cooling rates during thermal shock are about 6100 K/s and 5100 K/s, respectively. Additional dislocation semi-atomic surfaces (highlighted in yellow) form a strain field in IrNi nanoparticles; **b** dislocation motion is impeded by lattice distortion (red line) and dynamics of the millisecond quenching process

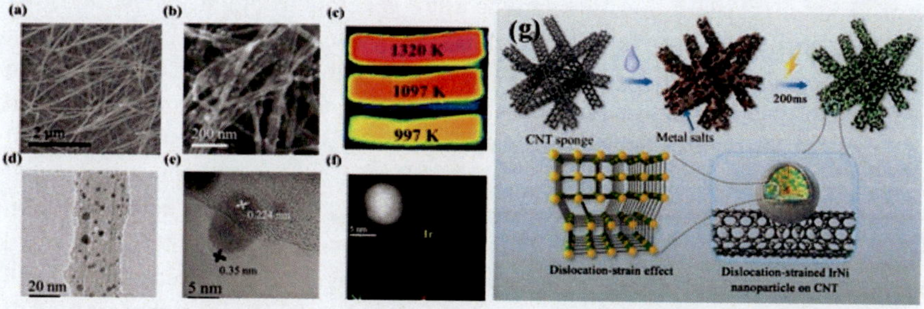

Fig. 2.3 **a** SEM image of CNTs loaded without metal salts; **b** SEM images of DSIrNi@CNTs; **c** a series of infrared temperature images showing the film heating process DSIrNi@CNTs as the applied electric power increases; **d** TEM images showed that IrNi nanoparticles were uniformly distributed on CNTs; **e** dislocation strain IrNi nanoparticles with thin carbon layers; **f** HAADF-STEM element mapping of Ir, Ni, and C at DSIrNi@CNTs; **g** schematic diagram of dislocation strain IrNi nanoparticles synthesis by HTS method

In addition, Yanan Chen's team also studied non-noble metal HER catalysts. A variety of materials based on the earth's abundant elements, for instance, MoS_2 [33–37], amorphous MoS_x [38, 39], MoC_2 [40], MoC_x [41], $MoSe_2$ [42], WS_2 [43], WC [44], WSe_2 [45], CoPS [46], $NiFeO_x$ [47], Ni_3S_2 [48], Ni_2P [49], $NiMoN_x$ [50], and

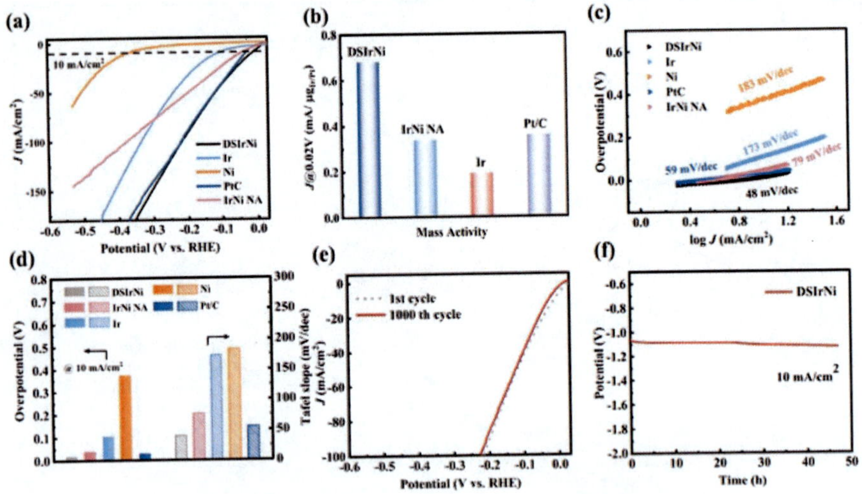

Fig. 2.4 **a** Curve of linear sweep voltammetry (LSV), measured at 5 mV/s in the aqueous solution of 1 M KOH; **b** the catalyst mass activity; **c** corresponding catalyst Tafel slope; **d** comparison of overpotential and Tafel slope; **e** polarization curve of HER after 1000 cycles; **f** HER timing potential curve at constant current density of 10 mA/cm^2

Ni-Mo [51] alloys are promising catalysts of HER. Among these abundant HER electrocatalysts on the earth, the cubic pyrite phase transition metal disulfide, for instance, ferric disulfide (pyrite, FeS_2), is becoming a new generation of highly active and low-cost substitutes [52]. Due to the rich content of pyrite, FeS_2 has been studied as a catalyst for the applications of energy conversion and storage [53, 54] widely, and FeS_2 which is nanostructured has been researched as an HER electrocatalyst with high activity. One work showed that utilizing film of FeS_2 as catalyst in solution of 0.5 M H_2SO_4 can achieve a current density of 10 mA/cm^2 at about overpotential of 260 mV [55]. FeS_2 films are synthesized by curing Fe films that handle electron beam evaporation. Despite advances in the research and design of cost-effective and efficient catalysts of HER, the methods of synthesizing nanostructured catalysts still need improving greatly to enable low-cost, rapid, and scalable nanomanufacturing of catalysts. Prof. Yanan Chen/Liangbing Hu prepared FeS_2-RGO (reduced graphene oxide) films using FeS_2 powder extracted from pyrite and GO (graphene oxide) flakes peeled from graphite by a modified Hummers approach as raw materials [56]. Primitive pyrite minerals are shown in Fig. 2.5a with samples a few centimeters long. FeS_2 particles of approximately 50 μm are obtained after using a hammer to crush the mineral. In order to better control the size of the particle in the experiment, commercial ultra-fine FeS_2 powder with particle size of about 44 μm and GO were prepared as raw materials. FeS_2-RGO film was prepared by vacuum filtration and annealing at the temperature of 573 K for 1 h in argon (Fig. 2.5b). Using the prepared FeS_2-RGO self-supporting film, the prepared FeS_2-RGO film was directly heated to high temperature by electric current-induced thermal shock, and treated at 2470 K for about 12 ms, and the nanosized FeS_2-RGO was in situ synthesized. After HTS treatment, 10–20 nm FeS_2 nanoparticles were distributed evenly on the nanosheets of RGO (Fig. 2.5c). The mechanism of ultra-fast in situ formation of FeS_2 nanoparticles on RGO at high temperature is that the FeS_2 powder is decomposed into atoms S and Fe during the process of rapid heating up to the temperature of 2470 K, and the atoms disperse in the matrix of RGO. Because the RGO is impermeable and the RGO film has encapsulation effect, Fe and S atoms can remain between RGO layers at high temperature. When rapid cooling occurs, the atoms S and Fe re-nucleate around the defects on the substrate surface of the RGO nanosheet and crystallize into ultra-fine FeS_2 nanoparticles. The 3D nanostructure of FeS_2-RGO contributes to keep great mechanical integration and fast electronic transmission of FeS_2 nanoparticles which were embedded in the RGO nanosheets. FeS_2 nanoparticles can be synthesized in situ by HTS method with ultra-fast, cost-effective, and scalable method. Thanks to the ultra-fine nanoparticles of FeS_2 and the robust 3D structure of FeS_2-RGO, the synthesized nanostructured FeS_2-RGO shows excellent HER electrocatalytic performance, reaching a current density of 10 mA/cm^2 with the overpotential of 139 mV in 0.5 M H_2SO_4 solution, and is capable of long-term operation. These results indicate that this is one of the most active electrocatalysts for HER reported in the essay around the world. The structure of ultra-fine nanoparticles, chemical composition, and their strong interaction with RGO enable them to have excellent electrocatalytic activity. This simple strategy of high temperature impact synthesis can be utilized to other transition metal binary, ternary, or multicomponent alloys and compounds, hence, stimulating new approaches to the development of crust-rich and highly active catalysts for various applications of scalable energy conversion.

2.3.2 Preparation of OER Catalyst by HTS

The electrochemical decomposition of water has been proven to be an economical, efficient, and clean technology of energy conversion, drawn great attention in the past few years. Due to its complicated four-electron transfer process, the slow kinetics of OER impedes the further development of this technology. Therefore, the development of efficient, durable, and low-cost electrocatalysts is essential to reduce overpotential and improve overall performance of OER [57, 58].

Ruthenium/iridium-based materials are the most advanced catalysts for OER. However, they are too few and costly for a wide range of utilization. To achieve economical and efficient OER, it is very important to develop catalyst materials which are precious metal free. Considerable efforts have been made to study catalysts based on transition metal, for example, sulfides, oxides of transition metal, phosphating compounds, and nitrides [59–61]. Among them, transition metal phosphates, such as Ni_2P, Co_2P, FeCoP, and CoFePx, are a kind of non-conventional precious metal materials with excellent OER electrocatalytic activity [62–64]. To further raise the catalysts activity without precious metals, extensive exploration has

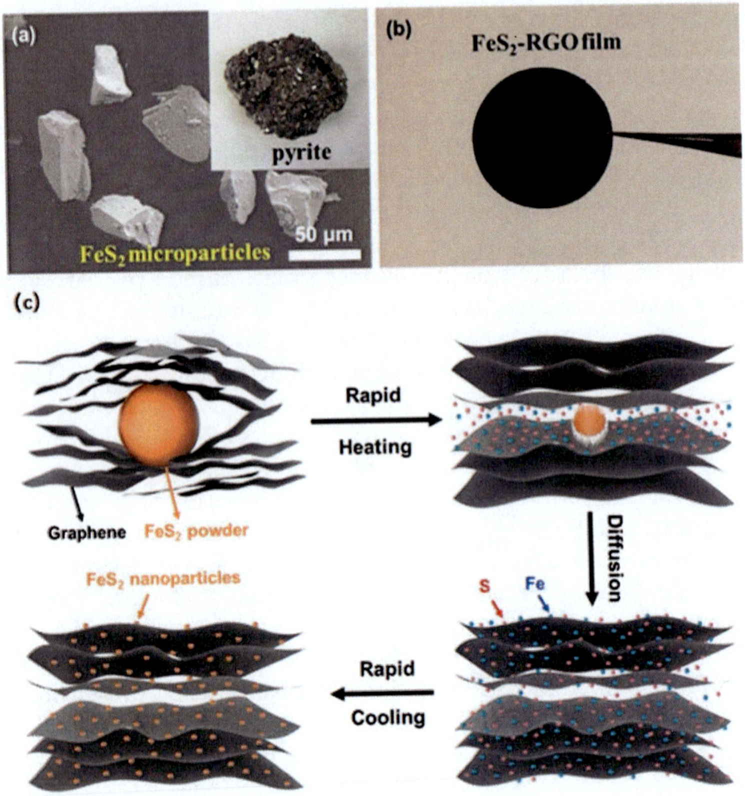

Fig. 2.5 **a** FESEM image of crushed FeS_2 powder from pyrite, inset presents original pyrite; **b** photographs of prepared FeS_2-RGO films; **c** schematic diagram of ultra-fast in situ transformation of minerals into catalyst nanoparticles

been carried out, and many materials based on transition metal have been designed to be sophisticated catalysts with high OER activity with nanostructures [65–68]. However, in the long-term reaction process, side reactions, for example, agglomeration, dissolution, corrosion, and inactivation of the nanocatalyst occur on the catalyst nanoparticles surface, thus reducing the catalytic stability and activity. As a result, it is very important to improve the nanocatalyst surface stability in OER process. Surface coatings have been shown to be an effective method to improve the surface stability of nanoparticles. However, designing coatings to protect the surface of catalysts without inactivating them is a great challenge. Carbon can modify the chemical stability of nanomaterials, which is one of the most widely used materials for coating [69]. But the carbon layer (particularly the thick layer) on the surface reduces the catalytic activity significantly. Using the method of coating to enhance the durability and avoiding sacrificing catalytic activity will help the overall catalysts performance greatly. In addition, using traditional methods to synthesize nanocatalysts is often complicated and time-consuming. Catalyst nanomaterials are prepared in a method of multi-step process and then cast onto conductive carriers such as "gas diffusion layers" utilizing polymer surfactants or binders, which results in weak catalyst-substrate adhesion and low conductivity [70]. Therefore, catalytically active coating nanocatalysts with long-term stability and high activity made by simple synthesis will be the best choice for OER catalysis.

Liangbing Hu's team designed a core–shell structure of cobalt iron phosphide ($CoFeP_x$) nanocatalyst with a thin iron shell (about 2 nm) as a protective layer for catalytic activity and demonstrated its high OER durability and activity [71]. $CoFeP_x$ nanoparticles with in situ Fe shells were synthesized by one-step HTS based on the difference of vapor pressure and instability of Fe and Co (Fig. 2.6). Carbonized wood (CW) is utilized as a substrate for the loading and synthesis of nanocatalysts. C-wood has great electrical conductivity and retains the natural wood structure, making it a perfect matrix for OER catalysts, providing 3D catalyst/substrate/electrolyte touch and low curvature channels of oxygen diffusion. The sample was subjected to HTS at about 1400 K in the ultra-fast time of 0.5 s, and the synthesis of $CoFeP_x$ nanoparticles was rapidly finished just in one step and firmly bonded to the channel wall of the C-wood. In this process, a thin layer of Fe is produced on $CoFeP_x$ nanoparticles in situ, and the stability of metal phosphide surface is enhanced during the process of OER, without sacrificing the catalytic activity because of the geometric effect of thin coating. As a non-noble metal catalyst, $CoFeP_x$ catalyst has high activity to OER in alkaline solution. The electrocatalytic oxygen evolution performance of $CoFeP_x$, CoP_x, and FeP_x on the substrate of CW is shown in Fig. 2.6d–g. At 10 mA/cm^2, the lowest initial potential of $CoFeP_x$ was about 1.55 V (vs RHE), i.e., the overpotential of $CoFeP_x$ was 323 mV, which was lower than that of FeP_x 418 mV and CoP_x 376 mV. What is more important is that the iron shell enhanced the $CoFeP_x$ nanocatalyst stability. The electrode potential of $CoFeP_x$ electrode can be stable at 1.57 V (vs RHE) for 50 h, showing excellent long-term durability. On the contrary, without shell protection, CoP_x and FeP_x nanoparticle electrodes showed poor stability performance, providing stability durations of about 25 h and 20 h respectively, suggesting that Fe shells have an advantage in improving the stability of phosphating catalysts. The synergistic effect of cobalt, iron, and phosphorus helps to improve the activity of OER, making it a low overpotential, high stability noble-metal-free catalyst.

HTS method can also be used to synthesize high-entropy compounds to catalyze OER reaction. Due to synergistic effects and high-entropy stability, high-entropy compounds usually perform greatly improved stability and catalytic activity [72–74]. Transition metal phosphates are widely used in biotransformation (e.g., vanadium oxyphosphate) [75], organic synthesis (e.g., zirconium phosphate) [76], oxygen evolution (e.g., cobalt phosphate, commonly known as CoPi) [77], and photocatalysis (e.g., silver phosphate) [78]. The polyanionic properties of these compounds, which include surface structures and unique electronic states, differ from the oxide catalysts used commonly or other catalysts, and they hold the promise of good selectivity and fast dynamics, or both. In order to synthesize high-entropy phosphate (HEPi), efficient decomposition of various precursors requires high temperatures, while short heating times are also critical to avoid element phase separation or segregation. But

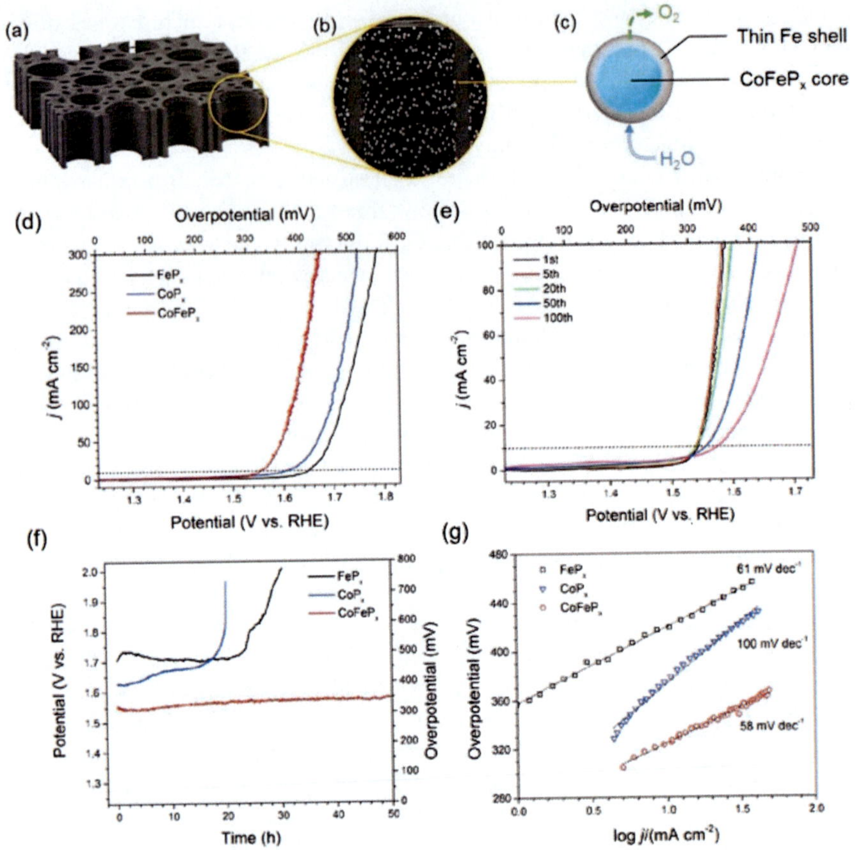

Fig. 2.6 **a** C-wood hierarchy with low curved duct channels and tracheids which enable oxygen diffusion; **b** $CoFeP_x$ nanoparticles grow in the C-wood channel with high catalyst loading; **c** schematic diagram of $CoFeP_x$ nanoparticles with thin iron shells; **d** polarization curves of $CoFeP_x$, FeP_x, and CoP_x in 1 M KOH at a scanning rate of 5 mV/s; **e** polarization curves of $CoFeP_x$ after different cycles of accelerated stability tests in 1 M KOH; **f** timing potential curves of $CoFeP_x$, FeP_x, and CoP_x recorded at 10 mA/cm^2 in 1 M KOH; **g** Tafel diagram of polarization curves of $CoFeP_x$, FeP_x, and CoP_x in 1 M KOH

these requests are beyond the capabilities of traditional synthetic methods and hinder the further development of HEPi materials.

Liangbing Hu's team reported the synthesis of the HEPi catalyst (CoFeNiMnMoPi), which is in the form of highly uniform spherical particles [79], as shown in Fig. 2.7. The synthesis method uses high temperature fly-through, an instantaneous high temperature process using aerosols to limit the uniform distribution of metal and phosphorus precursors in a single droplet. During rapid heating at high temperatures, the transition from oxide to phosphate can be performed in situ to make sure that the elements in the phosphate phase are evenly mixed within milliseconds. The team synthesized a series of high-entropy phosphate compounds, mono-metal, and trimetallic using HTS. In order to meet the different requests of catalytic reaction, the size of the particle should be well controlled by adjusting the synthesis parameters during the process (such as the concentration of precursor, flow rate, synthesis temperature, etc.). At the same time, spherical particles can also be made into hollow structures in situ because of the decomposition of trioctyl phosphine oxide (TOPO) as a phosphorus source. OER tests using HEPi particles as electrocatalysts in the model showed a much lower overpotential and faster kinetics than commercial IrO_x, with an overpotential of 270 mV at 10 mA/cm^2 and Tafel slope of 74 mV/dec. The outstanding performance owes to the high-entropy properties and synergistic effects of HEPi, as well as the high quality of the material brought about by the process of synthesis. This research sets up a new paradigm of synthesis for high-entropy polyanions for a series of utilizations in the catalytic and energy fields.

Transition metal sulfide (M_xS_y) catalysts have recently become candidate catalysts because of their excellent potentially good activity and electrical conductivity [80]. But the stability of M_xS_y is poor, including thermodynamic instability [81], structural and morphological deformation [82], and separation of catalyst from substrate under high oxidation electrochemical conditions [83]. In addition, monadic, binary, and ternary M_xS_y catalysts containing merely a small amount of metal elements lack extensive composition tunability. Conversely, according to Sabatier's principle

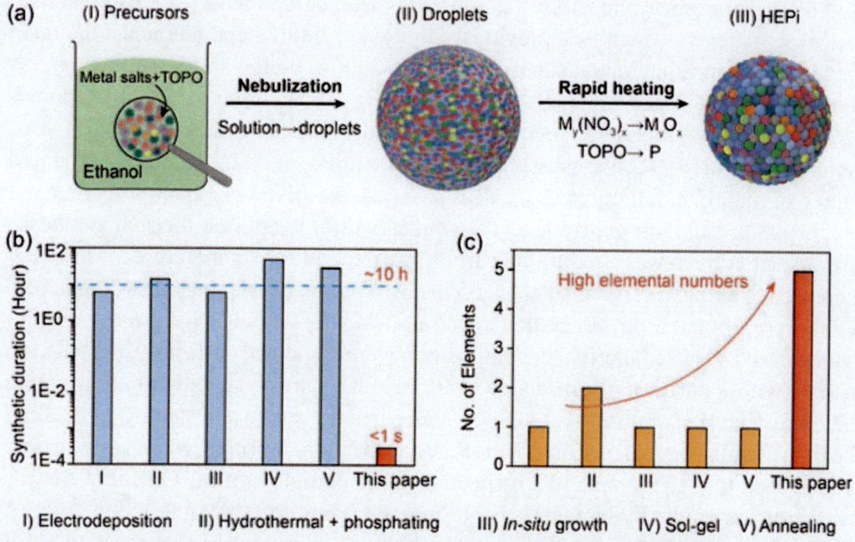

Fig. 2.7 Schematic diagram of the method of high temperature fly-through **a** schematic diagram of the formation process of HEPi particle: (I) TOPO and metal salt dissolved in ethanol to form mixed precursor solution; (II) aerosol droplets including metal salts and TOPO formed by atomization process in ethanol; (III) HEPi particles produced by rapid heating; **b**, **c** comparison of traditional and high temperature fly-through methods for the preparation of phosphate, including **b** synthesis time and **c** element quantity. *Note* I: electrodeposition, II: hydrothermal + hosphating, III: in situ growth, IV: sol–gel, V: annealing

[84], high-entropy metal sulfides (HEMS) with multiple metal elements have good compositive tunability to achieve optimal adsorption of reaction intermediates, thus further enhancing OER activity [85].

HEMS contains uniformly mixed polymetallic elements (≥5) in the sulfide structure. Due to the high-entropy properties of the stable phase, HEMS is expected to have good structural stability, which may improve the stability of OER. It is expected that the catalytic activity will be improved by adjusting the charge state of metal sulfide through multi-element coordination. Nevertheless, using a high-quality and efficient manner to synthesize HEMS remains a difficult task for the researchers. Up to now, only limited ternary and a few binary M_xS_y catalysts have been reported because of immiscibility limitations of incorporation of other elements. Bimetals $(NiCo)S_x$, $(FeCo)S_x$ and $(NiFe)S_x$ are not mutually soluble at most composition and temperature conditions, and it is particularly difficult to treat complex element compositions with different physicochemical properties using traditional methods. Besides, the synthesis of M_xS_y relies on time-consuming and energy-consuming multi-step processes heavily (such as post-vulcanization and pretreatment) [86], which does not allow for rapid screening and optimization. In addition, stabilization of solid solutions in relatively reactive sulfide structures (compared with oxides and carbides, etc.) presents extra challenges beyond the capabilities of conventional methods [87].

Liangbing Hu 's team synthesized M_xS_y nanoparticles for the first time by HTS method to overcome immiscibility of multiple metal components [88]. Compared to the M_xS_y material, which was previously limited to three metal elements, the team made an unprecedented advance in the synthesis of five-element HEMS nanoparticles, known as $(CrMnFeCoNi)S_x$, as shown in Fig. 2.8. In order to prepare nanoparticles of HEMS, a metal salt precursor (any kind of metal salt can be utilized to be the metal precursor, such as phosphate, acetate, nitride, or chloride, etc.) is used as a source of metal, and thiourea as a sulfur source is dissolved in ethanol together and supported on a carbon matrix (i.e., carbonized wood) for pulsed thermal synthesis. A transient current was used to the carbon matrix, and the temperature was rapidly increased to about 1650 K and kept at 55 ms. After rapid quenching, HEMS nanoparticles were obtained on the carbon substrate. The $(CrMnFeCoNi)S_x$ nanoparticles synthesized by pulsed thermal decomposition were dispersed uniformly and uniform in size, with a particle size of about 11.9 nm and a narrow size distribution within ±2.1 nm. The size of $(CrMnFeCoNi)S_x$ nanoparticles can be adjusted by controlling the time of pulse thermal decomposition. When the pulse thermal decomposition time is increased to 1 s, 5 s, and 10 s respectively, the particle size of $(CrMnFeCoNi)S_x$ nanoparticles gradually increases to 21.8 nm, 34.0 nm, and 40.9 nm.

The obtained $(CrMnFeCoNi)S_x$ nanoparticles were utilized as electrocatalysts for OER reaction in 1 M KOH solution and showed very low overpotential of 295 mV at 100 mA/cm^2 current density and great operational stability over 10 h. At 100 mA/cm^2, there was no serious overpotential increase. The HEMS nanoparticles' performance is superior to that of most M_xS_y materials which were reported in the previous articles, as well as its one-, binary, ternary, and quaternary M_xS_y counterparts. The super potentials of monadic, binary, ternary, and quaternary M_xS_y nanoparticles are negatively correlated with the amount of metal elements, showing that there is a

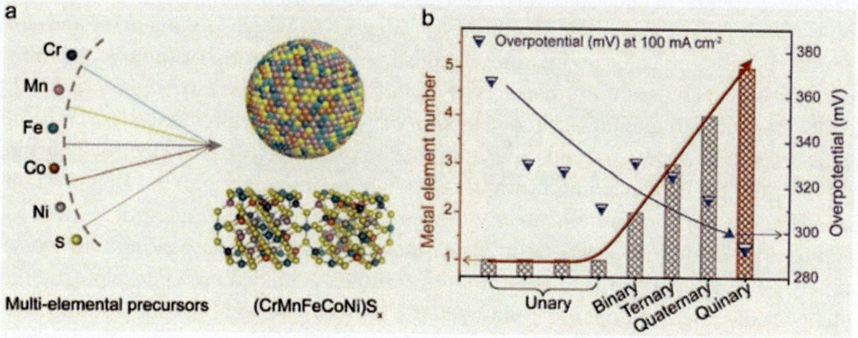

Fig. 2.8 Structure of HEMS (CrMnFeCoNi)S_x nanoparticles and their utilization as OER catalysts **a** immiscible metallic elements (i.e., Cr, Mn, Fe, Co, and Ni) are mixed to form homogeneous and high-entropy sulfide nanoparticles; **b** comparison of overpotentials and the amount of metal elements between monadic, binary, ternary, quaternary, and quintuple HEMS

synergistic effect between the metal elements. The excellent activity of OER and great durability of the synthesized HEMS suggest that it has a good potential as an electrocatalyst for efficient water decomposition. At the same time, the pulsed thermal decomposition method provides a general approach for the synthesis of multielement sulfides for a variety of applications.

2.3.3 Preparation of Dual-Function Catalyst by HTS

Pt, Ru, Ir, and other noble metal electrocatalysts have excellent HER and OER electrocatalytic activities. However, developing the noble metal electrocatalysts is limited by the low crustal abundance and high cost, and the traditional coating method of adhesives can increase resistance, bury the active site, and inhibit mass/electron transport. In addition, the load mass of electrocatalyst is usually less than 1 mg/cm^2, which provides limited catalytic active sites. As a result, a large number of studies has been devoted to develop the effective non-noble metal electrocatalysts and self-supported electrodes.

As a highly conductive substrate, nickel foam (NF) has a porous structure to offer an area with higher activity, and nickel has great electrocatalytic activity. Nanomaterials based on NF as self-supported electrocatalysts have been widely researched. Nickel foam is widely utilized in practical applications, for example, nickel metal hydride batteries, fuel cells, and electrocatalysts with mature industrial production and low cost. Recent researches have shown that nickel foam based nanostructured electrocatalysts are primarily prepared by solvothermal method or hydrothermal method. Nickel has high reaction activity with other additives under high pressure and high temperature, forming nanostructures (e.g., nanorods and nanosheets) in water or other organic solvents. But solvothermal and hydrothermal preparation request high pressure and temperature for a long time, which is dangerous and not conducive to large-scale production of industry. The utilization of organic solvents and additives increases costs and might be detrimental to the environment, which are not favorable to the industrial production of NF-based electrocatalysts. Hence, it needs to look for a low-cost and rapid approach to prepare NF-based electrocatalysts.

HTS provides a new method for ultra-fast nanomanufacturing. Yida Deng/Yanan Chen/Wenbin Hu's team innovatively designed and prepared an integrated electrocatalyst (NF-C/CoS/NiOOH) with NF as the substrate of joule heating and water immersion treatment [89]. Cobalt–thiourea complex is loaded on NF as a precursor, which is converted to doped carbon-coated CoS (NF-C/CoS) by joule heating and rapid cooling, thus enabling nickel activation into metastable, magically formed like a "seed", the "soil" fertilized by lightning (Fig. 2.9a). CoS/NF-C/NiOOH with nanosheet structure was prepared by easy immersion of CoS /NF-C in water, and the formation mechanism of nanostructure was put forward for the first time (Fig. 2.9b). Metastable nickel reacts with water to form NiOOH nanosheets, and the C/CoS in the "flowerpot" acts as a "seed" to induce spontaneous and continuous growth of the "flower-like" NiOOH nanosheets. These kinds of joule heating and immersion methods can quickly and simply synthesize nanomaterials with NF as self-supported electrocatalyst. Under joule heating condition, the cobalt–thiourea complex is converted into CoS with carbon-coated, which is N-, O-, S-doped carbon, and a small number of Ni is doped into CoS. It is noteworthy that, under joule heating and activation of cobalt–thiourea coordination complex, Ni on the surface of nanocrystalline spontaneously grows NiOOH nanosheets in water under CoS induction, thus producing NF-C/CoS/NiOOH nanosheets with core–shell heterostructure and hierarchical structure. It is speculated that the driving force for the formation of nanosheet structure is that Ni is in metastable state and CoS can induce the continuous growth of NiOOH nanosheet.

NF-C/CoS/NiOOH shows great electrocatalytic performance of HER, OER, and overall water decomposition. SEM, TEM, and electrochemical performance test images are shown in Fig. 2.10. When used as HER catalyst, the overpotential of naked NF and NF-C/Ni (OH) S at 10 mA/cm^2 current density is 347 mV and 221 mV, respectively, while NF-C/CoS/NiOOH has 170 mV overpotential. NF-C/CoS/NiOOH still has a low overpotential of 294 mV at a high current density of 100 mA/cm^2. In addition, the R_{ct} of the EIS test for NF-C/CoS/NiOOH was 12.3 ω, and the Tafel slope was 120.96 mV/dec, indicating that the synthesized NF-C/CoS/NiOOH had high electron transfer speed, superior HER reaction kinetics, and higher density of catalytic active sites. When used as OER catalyst, the overpotential of bare NF and NF-C/Ni(OH)S is 422 mV and 319 mV respectively at the current density of 10 mA/cm^2, while NF-C/CoS/NiOOH has 296 mV, showing good OER performance. At a high current density of 100 mA/cm^2, NF-C/CoS/NiOOH has a low overpotential of 361 mV with a R_{ct} of 3.5 ω and a Tafel slope of 52.9 mV/dec, indicating that NF-C/CoS/NiOOH has a low electron transport barrier, high charge transfer coefficient, and superior kinetics of OER reaction.

Transition metal chalcogenides, oxides, and other transition metal compounds (nitrides, borides, carbides, phosphates, etc.) have also been widely studied for

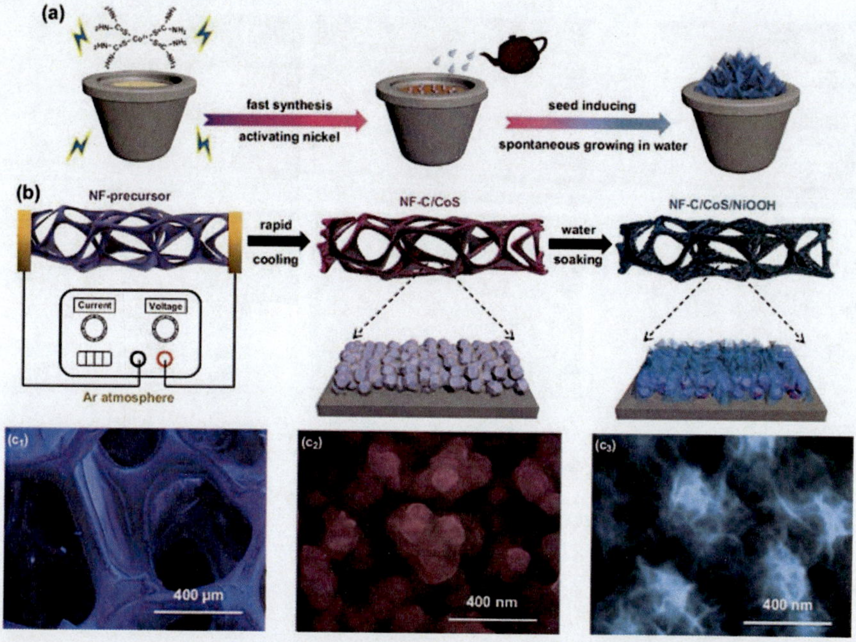

Fig. 2.9 **a** CoS/NF-C/NiOOH synthesis strategy; **b** schematic diagram of NF-C/CoS/NiOOH preparation; SEM images of **c1** NF- precursor, **c2** NF-C/CoS, and **c3** NF-NF-C/CoS/NiOOH

their abundance, low price, and good electrochemical properties. In 2017, Shaomao Xu/Yanan Chen/Liangbing Hu reported that Co_2B nanocatalyst can be directly produced from large Co_2B body precursors by HTS method [90], and the synthesis diagram is shown in Fig. 2.11. Co_2B precursor was produced by the reaction of sodium borohydride ($NaBH_4$) with cobalt acetate (CoA_{c2}) solution. Co_2B particles were mixed with GO solution under continuous agitation, and then the mixture was cast onto a flat slide and prereduced to form Co_2B/RGO composite film. Co_2B blocks were dispersed into the RGO film uniformly. The massive Co_2B precursor (micron size) can be converted into ultra-fine nanoparticles with diameters of 10–20 nm by the treatment of continuous high-temperature pulse at about 1900 K for 1 s in an argon atmosphere. After high temperature treatment, nanosized Co_2B catalyst was uniformly distributed on RGO sheet. The prepared Co_2B/RGO composite showed great electrochemical properties of water decomposition and excellent cyclic stability. In addition, the team succeeded to prepare cobalt oxide (Co_3O_4) and molybdenum disulfide (MoS_2) nanoparticles by high temperature pulses.

Co_2B showed excellent dual-functional catalytic performance of HER and OER. For testing the activity of HER, the CO_2B-modified electrode acted as a cathode and had an initial overpotential as low as 33 mV under acidic conditions. The low overpotential of 260 mV at a high current density of 100 mA/cm^2 makes the synthesized Co_2B nanoparticles one of the most efficient catalysts for HER. Co_2B nanoparticles have excellent stability in long-term operation, and as a stable catalyst for HER, they

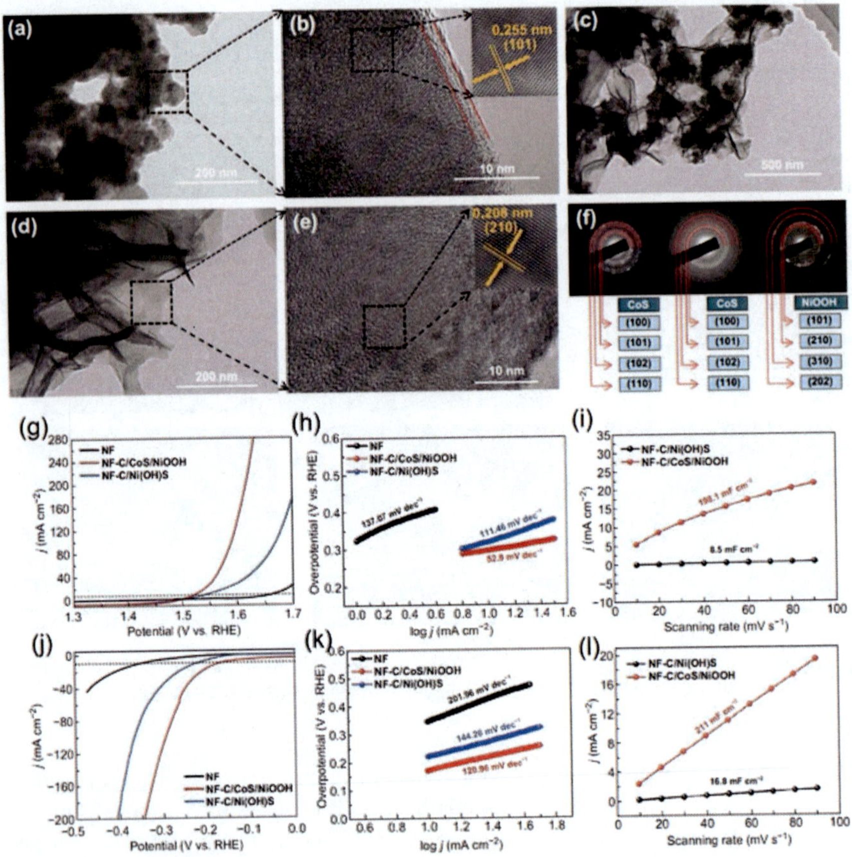

Fig. 2.10 **a** TEM of NF-C/CoS; **b** HRTEM of NF-C/CoS (illustration: FFT inverse mode); **c–d** CoS/NF-C/NiOOH SEM; **e** HRTEM of CoS/NF-C/NiOOH (illustration: FFT inverse); **f** SAED models of **a**, **c**, and **d** respectively. **g** Polarization curves and **h** Tafel slope of naked NF, NF-C/CoS/NiOOH, and NF-C/Ni (OH)S in OER test; **i** C_{dl} of NF-C/CoS/NiOOH and NF-C/Ni(OH)S in OER; **j** polarization curves and **k** Tafel slope on naked NF, NF-C/CoS/NiOOH, and NF-C/Ni(OH)S in HER test; (l) C_{dl} of NF-C/CoS/NiOOH and NF-C/Ni(OH)S in HER

can be used for long-term applications under high workloads. When the OER reaction was catalyzed by Co_2B nanoparticles as an anode, the initial potential of the oxygen evolution reaction was 1.59 V (vs RHE), corresponding to an initial overpotential as low as 360 mV. The calculated slope of Tafel is exceeding low, 85.4 mV/dec, with equivalent good OER activity.

Recently, materials based on cobalt sulfides (e.g., cobalt selenide and cobalt sulfide) are emerging as promising efficient and low-cost HER or OER catalysts. However, there are few reports on the use of cobalt sulfide as a single dual-function catalyst to decompose water in integrated electrolytic cells. Besides, the

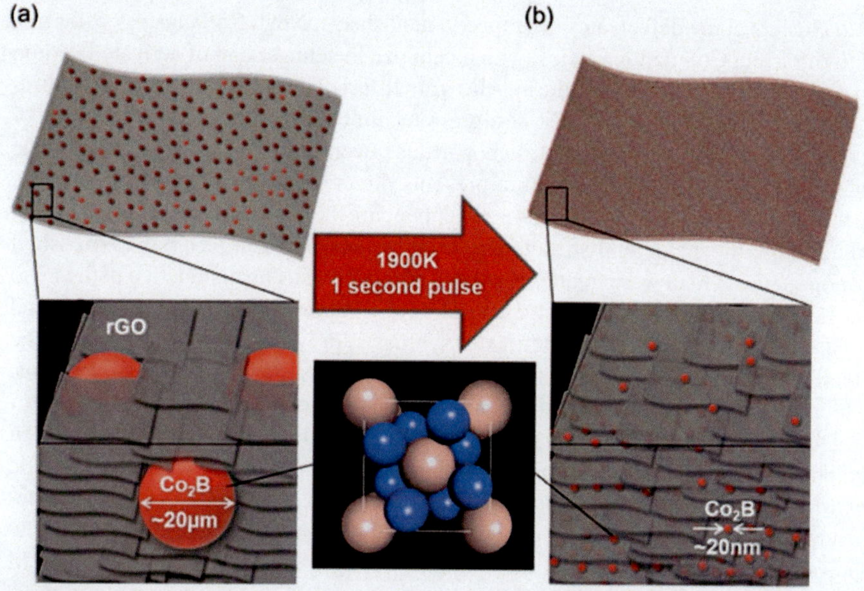

Fig. 2.11 Schematic diagram of Co_2B nanoparticles synthesized by new high temperature pulse processing **a** After the chemical reaction between $NaBH_4$ and CoA_{c2}, Co_2B micron particles are encapsulated in RGO films. **b** After high temperature pulse process, Co_2B particles were transformed into ultra-fine nanoparticles with a diameter of about 20 nm, which were modified on RGO sheets uniformly. The inset presents the Co_2B crystal structure, which remains unchanged after the high temperature pulse

most common method to synthesize these catalysts is hydrothermal method in solution, which costs a lot of time and is difficult to a large-scale production. In addition, electrodeposition of cobalt sulfide has been reported to catalyze HER. Nevertheless, this method often makes catalyst particles aggregate and weak interaction with substrates, which will result in poor performance of catalysis. Catalysts which were synthesized by the above approaches are always exposed to the electrolyte directly, which is usually a strongly acidic or alkaline solution, resulting in the unavoidable corrosion of the active material in the electrolytic cell.

Yanan Chen/Liangbing Hu reported the rapid synthesis method of CoS@few-layer graphene core–shell electrocatalysts by HTS method for efficient whole water cracking [91]. In Fig. 2.12, the synthesis of core–shell electrocatalyst of cobalt sulfide @ few graphene layers is schematically described. Films loaded with precursors were prepared by mixing thiourea, cobalt acetate, and GO (Fig. 2.12a) and then pre-annealed at 673 K in a tubular furnace. The original film was changed into RGO nanocomposites and cobalt sulfide nanoparticles by applying a HTS of 2000 K under the current pulse. During in situ synthesis, cobalt sulfide nanoparticles are coated with several graphene shells. The formation mechanism of core–shell electrocatalyst of cobalt sulfide @ few graphene layers is proposed as follows: the high temperature caused by HTS generates the pyrolysis of metal salts, and then CoS recrystallizes into clusters at the defect sites of graphene nanosheets. Next, CoS clusters will grow into ultra-fine CoS nanoparticles, which catalyze the conversion of activated carbon to graphene on CoS surface during the growth procedure (Fig. 2.12b). The ultra-short impact time and ultra-fast cooling rate limit the migration and diffusion of CoS nanoparticles and make the nanoparticles evenly distributed on the graphene nanosheet. This core–shell structure prevents direct exposure of the catalyst to the electrolyte, thus restricting corrosion and enhancing the electrocatalytic performance. In addition, the stable and strong interfacial contact between RGO and core–shell nanoparticles which were treated at 2000 K provides a strong conductive pathway to further improve catalytic performance. The RGO nanosheets are doped with N and S heteroatoms in thiourea during in situ synthesis, providing extra active sites. The stabilization and formation of nanoparticles in the matrix took place in about 7 ms, which was better than the traditional manufacturing method.

Due to the highly conductive graphene coating and the strong interaction between graphene and nanoparticles, the core–shell nanocomposites show great catalytic performance for overall hydrolyzing. In a typical three-electrode battery, an independent RGO-CoS film was set up to be utilized to be the working electrode directly to evaluate the HER activity of the synthetic catalyst, and the electrolyte was 1 M KOH solution. CoS-RGO showed great activity for HER, and the lowest initial overpotential was about 37 mV compared with reversible hydrogen electrode (RHE). CoS-RGO electrocatalysts require 51 mV, 118 mV, and 226 mV to reach the cathode current densities of -1 mA/cm^2, -10 mA/cm^2, and -100 mA/cm^2, respectively. The Tafel slope of CoS-RGO is about 63 mV/dec, higher than 51 mV/dec of 40 wt.% Pt/C

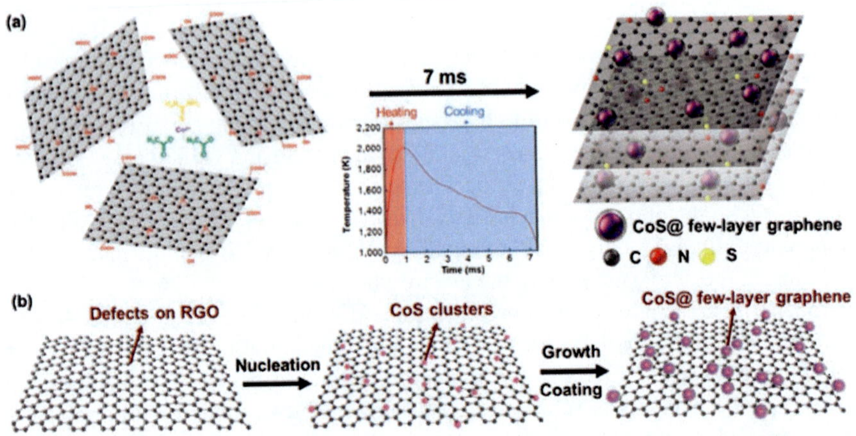

Fig. 2.12 Synthesis of core–shell nanocatalysts of cobalt sulfide @ few graphene layers in milliseconds **a** Cobalt acetate, thiourea, and GO were transformed into ultra-fine uniformly loaded cobalt sulfide @ few-layer graphene core–shell nanoparticles after HTS at about 2000 K, and the graphene was doped with nitrogen and sulfur. The curve shows the temperature distribution during high temperature treatment; **b** the formation mechanism of cobalt @ few layers' graphene

electrocatalyst, making it one of the most active non-precious metal HER electrocatalysts under alkaline conditions. CoS-RGO electrocatalyst's OER activity was also studied in 1 M KOH solution. Linear scanning of CoS-RGO shows that, compared with RHE, a small potential of 1.58 V and 1.66 V is needed to drive 10 mA/cm^2 and 100 mA/cm^2 cathodic currents, compared to Ir/C (20 wt.% Ir), showing that CoS-RGO has excellent OER electrocatalytic performance. In contrast, the current exhibited by pure RGO in the range of potential measured is negligible, indicating that the contribution of pure RGO to OER performance is negligible. The Tafel slope of about 71 mV/dec proves the great catalytic property of CoS-RGO for OER, which is superior to the Tafel slope of 118 mV/dec for Pt/C catalyst reported. In order to study the feasibility of using CoS-RGO as an effective dual-function catalyst for both HER and OER, CoS-RGO-independent electrodes were fabricated as cathode and anode in 1 M KOH solution in a dual-electrode battery device. CoS-RGO requires about 1.75 V and 1.97 V anodic currents of 10 mA/cm^2 and 100 mA/cm^2, respectively, which are superior to the Pt electrocatalysts reported. The stability test of CoS-RGO showed no attenuation of stability at different current densities. In contrast, the CoS nanoparticles synthesized on the carbon fiber (CF) network without graphene indicated significant attenuation of HER activity after merely three cycles, indicating that CoS-CF without the protection of graphene was less stable. The experimental results show that this method can be widely used to synthesize other transition metal chalcogenides with several layers of transition metal alloys, transition metals, graphene coatings, and other nanoscale composites.

2.4 Preparation of Fuel Cell Catalyst by HTS

Fuel cell (FC) is a device of electrochemical reaction that converts the chemical energy of fuel and oxidizer into electrical energy directly. It consists of a cathode, an anode, and an electrolyte membrane. The oxidant is reduced at the cathode, and the fuel is oxidized at the anode to complete the electrochemical reaction. The cathode side is connected to oxygen or air, and the anode side is connected to different fuels (such as hydrogen, methane, gas, etc.), which can be divided into hydrogen fuel cell and hydrocarbon fuel cell. Common fuel cells include direct methanol fuel cell (DMFC) and proton-exchange membrane fuel cell (PEMFC) developed on this basis. Table 2.1 shows the characteristic states of some fuel cells.

Fuel cells have the advantages of low material cost, wide fuel adaptability, high power density, high conversion rate from chemical energy to electric energy, and environmental friendliness. However, there are still many shortcomings of fuel cells, which restrict the large-scale application and commercialization of fuel cells. The bottleneck of application includes high manufacturing cost, unsatisfactory durability, life, and stability at high temperature, and technical problems such as catalyst poisoning and deactivation. Therefore, the synthesis of cheap, stable, and efficient battery electrode catalyst is an important way to solve the problem of fuel cell development.

Table 2.1 Characteristic states of various fuel cells

	AFC	PAFC	PEMFC	MCFC	SOFC
Electrolyte	KOH	H_3PO_4	Proton-exchange membrane	Li_2CO_3, Na_2CO_3	Ceramic
Conductive ion	OH^-	H^+	H^+	CO_3^{2-}	O^{2-}
Working temperature/°C	50–200	180–200	Room temperature ~150	600–700	600–1000
Fuel	Pure hydrogen	Reforming hydrogen	Reforming hydrogen	Hydrogen, natural gas, biofuels	Hydrogen, natural gas, biofuels
Oxidant	Pure oxygen	Air	Air	Air	Air

2.4.1 Preparation of Cathode Catalyst by HTS

As the demand for sustainable energy supply increases, metal-air cells and fuel cells have drawn extremely important attention because of their renewable fuel sources, high energy density, and great environmental friendliness [92]. One of the big challenges of these kinds of batteries is the development of the air electrodes with high efficiency of energy conversion [93]. The ORR at the air cathode is important for the electrochemical energy conversion from oxygen to water during the discharge of fuel cells or metal-air cells. ORR occurs at the sites of three-phase boundary generally, including aqueous electrolyte, gaseous oxygen, and solid catalyst, and the active site of the ORR depends on the catalyst and carrier in the air cathode [94]. On the one hand, it needs stable ORR catalyst, which could effectively accelerate the slow oxygen reduction kinetics. On the other hand, it also requires porous conductive carriers (containing gas diffusion electrodes and catalyst carriers) which can offer anchor catalyst locations and interconnect porous structures for charge and mass transfer. Preparing this kind of integrated high-performance air cathodes directly still remains some challenges due to their complexity. Traditional Pt/C catalysts have been utilized widely and are regarded as one of the most effective catalysts for ORR [92]. The poor stability, high cost, and toxicity of some intermediates limit their widespread application in metal-air cells or fuel cells. Alloying of precious metals with transition metals (e.g., Fe, Co, Ni, Ag, and Ce) has been widely used in the design of catalysts with lower amounts of precious metals and improved catalytic performance. The incorporation of 3D transition metals can change the electronic structure of Pt and shorten the bond of Pt–Pt, thus facilitating the dissociation and adsorption of oxygen [95–97]. In the preparation of conventional air electrodes, the powdered catalyst is usually supported on a carbon carrier and then coated on a porous gas diffusion layer with the assist of a polymer binder. Although this approach is feasible and widely utilized, there are several problems with the prepared air electrode that can lead to reduced catalytic performance: Firstly, during the long-term operation of acid/base electrolytes, it is common for catalysts to fall off from the current collector. In order to maintain catalytic activity, a large number of catalysts are inevitably needed. Secondly, because of the inert surface of the carbon support, the interaction between the catalyst and the carbon support is very weak, which leads to seriously agglomeration or migration of metal nanoparticles, which reduces the active surface area and results in the degradation of the performance during long-term use. Thirdly, the use of polymer adhesives increases the internal resistance of the battery and covers the catalyst to reduce catalytic activity. Hence, air electrodes with low platinum content, strong catalyst–carrier bonding, but stable and high catalytic activity are urgently needed.

Qingwen Li's team pyrolyzed ferrocene (growth catalyst for carbon nanotubes) to obtain iron nanoparticles and distributed them in a highly conductive carbon nanotube (CNT) network [98]. From the perspective of air electrode design, this supported catalyst structure is a selection with great promise. But CNT films loaded with iron nanoparticles presented poor activity of intrinsic ORR. Therefore, Hu's team utilized this special structure to achieve the production of high-performance air electrodes by prioritizing iron nanoparticles alloyed with platinum in interconnected CNT films. Figure 2.13a shows the manufacture of an air cathode for interconnecting the CNT network with well-dispersed PtFe alloy nanoparticles anchored to the CNT. Firstly, the prepared CNTs films were electrochemical oxidized to improve their wettability. Then, the films were immersed in aqueous solution of H_2PtCl_6, and the part of Fe reduced Pt^{4+} to Pt, which was attached to the iron particles. Finally, the film was heated by transient joule heating, leading to the generation of well-alloyed PtFe nanoparticles in the film. On account of retaining the unique features of carbon nanotube film, for example, high mechanical strength, high electrical conductivity, and porous network, the film can be obtained as a highly integrated air cathode directly, using carbon nanotube film as the current collector, no glue and gas diffusion layer of catalyst carrier, PtFe alloy nanoparticles can be used as a catalyst for ORR. Compared with unprocessed CNT films, DCNT films using electrochemical etching showed enhanced ORR activity (Fig. 2.13b–g), with a corrected initial potential and an improved half-wave potential, indicating that the existing of defects in CNT can improve catalytic activity of ORR. The PtFe-DCNT films with low Pt content of 1.7 wt.% show catalytic property of ORR, with an initial potential of 0.95 V (vs RHE), a half-wave potential of 0.84 V (vs RHE), and a current density of 5.3 mA/cm^2, compared to commercial Pt/C catalysts with a higher Pt load (20 wt.%). PtFe-DCNT films provide a considerable mass activity of ORR of 0.33 A/mg at 0.85 V (vs RHE), which is more than six times that of commercial Pt/C catalysts, with outstanding high tolerance and long-term stability to methanol.

Although nanoparticles of various shapes and sizes have been widely studied for a variety of catalytic applications, most studied nanoparticles contain only three or fewer elements to prevent structural heterogeneity and synthetic complexity [99, 100]. Multi-metallic nanoclusters (MMNCs, superfine nanoparticles with more than three elements) represent a huge and undiscovered space of chemistry which promises to fine-tune material properties through high-dimensional compositional control and related synergistic interactions between different elements [101]. With

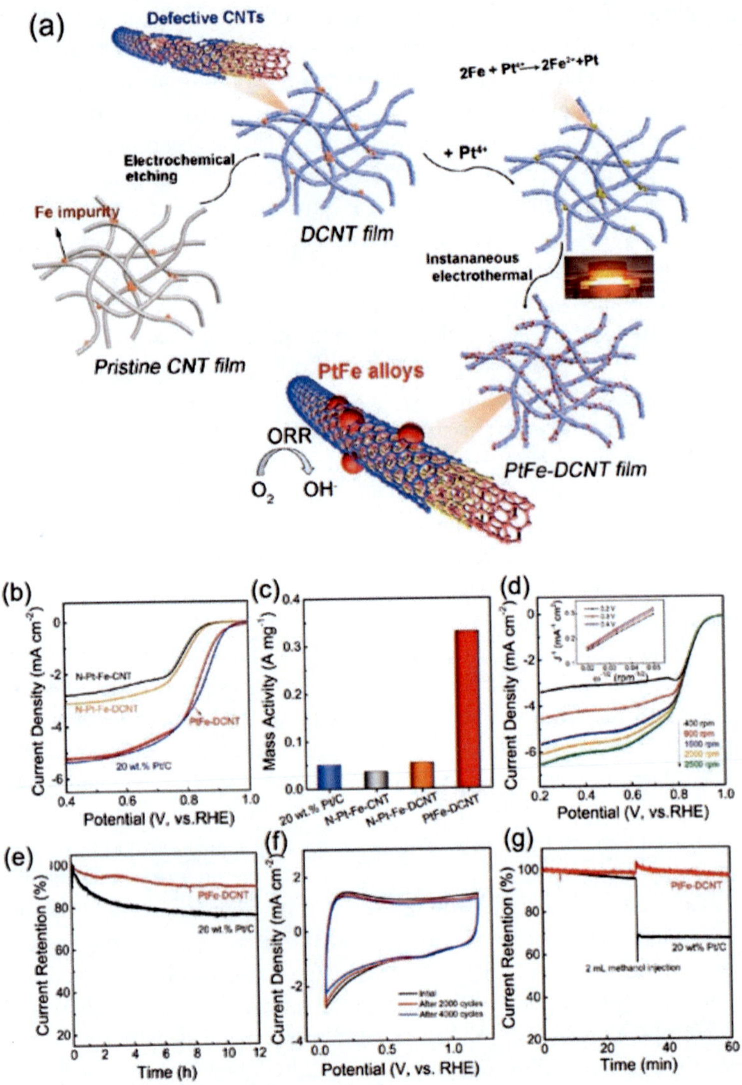

Fig. 2.13 **a** Schematic diagram of the manufacturing procedure of PtFe-DCNT film; **b** the ORR LSV curves of the catalyst films of PtFe-DCNT, N-Pt-Fe-DCNT, N-Pt-Fe-CNT, and 20 wt.% Pt/C in oxygen-saturated 0.1 M KOH aqueous electrolyte at 1600 rpm; **c** the mass activity of PtFe-DCNT, N-Pt-Fe-DCNT, N-Pt-Fe-CNT, and 20 wt.% Pt/C catalyst films; **d** ORR LSV curves of PtFe-DCNT films at different rates (400–2500 rpm) (illustration: KL diagram of PtFe-DCNT films); **e** comparison of the duration of catalyst films with PtFe and 20 wt.% Pt/C at 0.5 V vs RHE; **f** accelerated degradation test of PtFe-DCNT catalyst film: CV curves after 2000 cycles and 4000 cycles, scanning rate of 200 mV/s, showing outstanding durability of catalyst film; **g** methanol antitoxicity of PtFe-DCNT and 20 wt.% Pt/C catalyst

the complexity of the composition increasing, traditional synthesis methods typically result in a wider size distribution and uneven structure of MMNCs (e.g., phase separation and/or element separation within the particle) [102]. These inhomogeneity result from the enormous challenge of controlling the thermodynamics and dynamics of chemical reactions between these different components at the nanoscale. Structural and size heterogeneity make it difficult to control composition adjustments and systematically study MMNCs, restricting the discovery of material, performance optimization, and mechanical understanding of different functions. Besides, with the number of components increasing, the combination of the resulting explosion in synthetic space will need a lot of investment on MMNCs synthesis and screening, this poses a huge challenge to the traditional methods, the traditional method is usually very slow (a sample a day or less) and component specific (i.e., the processing parameters are usually does not apply to different ingredients). Compared to sample-by-sample trial and error, a high-throughput paradigm can offer parallel synthesis of many samples with different components, which saves both effort and time. Pioneering work has proved that combinatorial synthesis of a variety of heterogeneous catalysts (with three to five elements) synthesized utilizing techniques of thin film deposition or space-constrained tips.

Liangbing Hu's team reported high-throughput synthesis and screening of a sequence of alloyed and ultra-fine MMNCs which were supported on surface-treated carbon carriers [103] in the PtPdRhRuIrFeCoNi space, as shown in Fig. 2.14. High-throughput synthesis is achieved by combining composition formulations in a solution phase followed by rapid HTS treatment. During this process, the defects on the carbon surface contribute to disperse the MMNCs and guarantee its dimensional uniformity in the samples with different compositions, while the rapid impact procedure results in structure of single-phase because of high temperature mixing and rapid quenching. These MMNCs with different compositions (of similar structure and size) were screened fast by electrochemical oxygen reduction reaction (ORR) utilizing scanning droplet cell analysis, making us quickly identify two catalysts with the best performance. To confirm the high performance of two optimized catalysts (PtPdFeCoNi and PtPdRhNi), electrochemical analysis was performed and compared with Pt samples (10 wt.% supported). Compared with Pt, MMNCs catalyst has lower overpotential and better catalytic activity. In addition, the peak current density of the two MMNCs catalysts increased by about twice as much as the peak current density of Pt, with higher limiting current density and positive half-wave potential, further proving the better activity of the two MMNCs samples compared to Pt controls. The long-term stability of the prepared catalysts was tested at 0.6 V (vs RHE) by timing current method. The current density of PtPdRhNi reduced by 36% after running 15 h, and the activity of PtPdFeCoNi reduced by 29% after running 15 h. In comparison, the current density of Pt reduced by 39.1% after the same process.

The method of high-throughput synthesis and screening provides a way to speed up the composition probe of MMNCs research, lay the foundation of the fast synthesis and composition probe of advanced catalytic materials, and provides machine learning and data mining inspiration for development of MMNCs catalysts in the future.

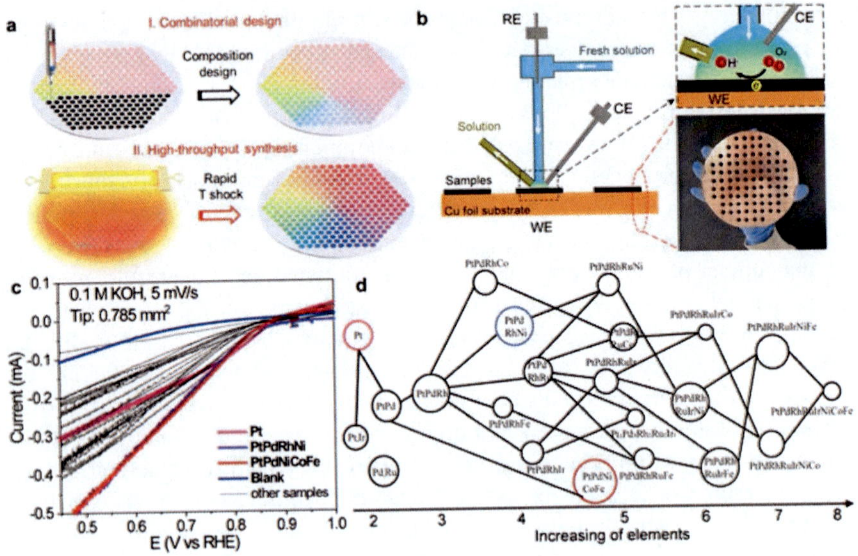

Fig. 2.14 Magnified synthesis and rapid screening of MMNCs for electrocatalytic reaction **a** schematic diagram of uniform MMNCs combination and high-throughput synthesis; **b** scanning droplet unit setup and patterned samples on a copper substrate. CE, contrast electrode; RE, reference electrode; WE, working electrode; **c** rapid screening of PtPd-based MMNCs for catalytic ORR (22 components +1 blank, 0.1 M KOH, scanning rate: 5 mV S^{-1}). **d** the combination design presented in the neural network diagram and its corresponding ORR performance. The size of the circle represents the specific current size of the ORR at 0.45 V

Dual-functional electrocatalysts that facilitate ORR or OER are significant for rechargeable metal-air batteries development with low-cost, high-energy density, and environmental friendliness. Although RuO_2/IrO_2 catalysts for OER and Pt-based catalysts for ORR are considered the most advanced oxygen catalysts, they still suffer from high cost, lacking dual-functional catalytic activity and poor stability of alkaline electrolytes for metal-air batteries. Lots of types of dual-function electrocatalysts have been found that can operate at low overpotential in alkaline electrolytes. They are primarily based on transition metal compounds or alloys (such as Co_3O_4, $NiCo_2O_4$, $ZnCo_2O_4$, CoP, NiCo, NiFe, and FeCoNi) and doped carbon nanomaterials (e.g., N, B, P, S dopants). Therefore, methods like heteroatomic doping, anion and cation regulation, strain engineering, and defect production are suitable for regulating the catalysts' electronic structure and generating lots of active sites for ORR and OER. Because ORR and OER occur at three-phase boundary sites involving gaseous oxygen, aqueous electrolytes, and solid catalyst surfaces. Effective electrocatalysts require the integration of adequate OER and ORR active sites as well as appropriate carriers to firmly anchor the catalyst and provide mass and charge transport paths. A few integrated catalysts, such as nickel foam, CNT foam, and carbon cloth, have already been synthesized by loading transition metal-based catalysts in a 3D conductive network. These catalysts usually do not contain adhesives and can be used directly as air cathodes. However, in order to achieve acceptable performance of OER and ORR, large loads of such catalysts are required. Disadvantages like increased weight of battery (low energy density), blocking mass transfer paths, and catalyst spalling during long periods of operation may be unavoidable. In addition, due to the limitations of the catalyst loading methods currently in use (such as sol–gel approaches), the control of catalyst size and the reservation of the 3D structure of the carrier during catalyst loading remains challenging.

Qingwen Li's team reported fast preparation of CoNi alloy dual-functional catalysts supported inside CNT networks by HTS method [104]. CoNi nanoparticles (CoNi@N-DCNT) were uniformly anchored in CNT membranes by treating N-DCNT membranes permeated by Ni^{2+} and Co^{2+} precursors by transient joule heating. Although the load mass of CoNi nanoparticles is only about 0.06 mg/cm^2, 13 wt.% of the entire composite film, CoNi@N-DCNT film shows high bifocal catalytic performance and excellent long-term stability for OER and ORR when used as an integrated air electrode. In addition, the team used rapid pulses of high temperature to immediately break down a group of metal precursors which contains 10% precious metal atoms (platinum and iridium) and 90% transition metal atoms (iron, cobalt, and nickel), which were subsequently solidified into multi-element nanoparticles ($FeCoNiO_x$). Precious metals form ultra-small nanoparticles (IrPt, about 5 nm in diameter) anchored on the surface of transition metal oxide nanoparticles, forming a layered structure. This hierarchical nanostructure reduces the use of noble metals, produces a huge electrically active surface area, and stabilizes IrPt alloys in the process of the oxygen electrode reactions. At the same overpotential, the mass catalytic activity of the IrPt nanoparticles was 7 times higher than that of Pt in ORR and 28 times higher than that of Ir in OER, indicating the bifocal electrocatalyst has high activity. This excellent performance is due to the controlled composition of multiple elements, large electrically active surface area, and mixed chemical states. The multi-element design and layered nanostructure of these nanoparticles provide a versatile and powerful replaceable material for solar batteries, catalysis, etc.

2.4.2 Preparation of Anode Catalyst by HTS

Alkaline direct alcohol fuel cells (ADAFC) fueled by ethylene glycol, ethanol, and methanol are regarded as an emerging electrochemical device which can convert chemical energy into electrical energy through liquid fuel [105]. Alcohols with low molecular weight, such as ethanol, are economical, clean, and sustainable sources of energy that are easy to obtain, store, and transport. Up to now, catalysts of platinum-based precious metal are thought to be efficient electrocatalysts for the reaction of alkaline alcohol oxidation, but they are vulnerable to CO poisoning, slow anodic oxidation reaction kinetics, and high precious metal price seriously hinder their development. Therefore, the development of low-cost and high efficiency anode electrolytic capacitors has become the focus of large-scale commercialization of fuel cells in the future. Non-precious metal materials are considered as proper and promising modified electrocatalysts for the electrooxidation of alcohols. Nickel-based materials have been introduced as active electrocatalysts for the oxidation of methanol and other alcohols. Although considerable progress has been made, the high initial potential and low catalytic activity of alcohol electrooxidation seriously block the further application of nickel-based materials. Therefore, it is essential to design catalysts with novel structure to improve the oxidation activity of alkali alcohols. Defect engineering of nanomaterials is an effective strategy to regulate the electronic structure and crystal structure of surface atoms and to introduce a variety of defect sites as active centers. Among them, oxygen vacancy (V_O) can enhance the catalytic activity through optimizing the electronic structure of metal atoms, which can enhance the electrical conductivity and increase the reactivity and number of transition metal oxide materials. In addition, V_O in the catalyst can facilitate the charge transfer between the reactant molecules and the catalyst, thus improving the catalyst adsorption capacity [106]. It is believed that the deep structural modulation of electrocatalysts under controlled instability will result in metastable phase and/or incomplete structural defects. Under such circumstance, the imperfect structure in the high-energy state can provide many active sites for the electrocatalytic reaction, thus improving the catalytic activity. Therefore, the development of nickel-based materials with sufficient defect sites is a potential way to realize the production of alcohol electrooxidation catalysis.

Yanan Chen/Yida Deng/Wenbin Hu 's team proposed the rapid synthesis of highly dispersed nickel oxide (NiO) nanoparticles (NiO@C/CC), which were coated with ultra-thin carbon layer on carbon cloth by HTS method. This synthetic material is rich in high-valence Ni species and oxygen vacancies, and the synthesis route is shown in Fig. 2.15 [107].

HTS can drive rapid processes of heating and quenching, provide an energy barrier for steady-state atomic movement and rearrangement, and induce defective and higher energy oxygen atoms, known as metastable states. Through the synergistic effect between oxygen-rich vacancy, high concentration of Ni^{3+} and ultra-thin carbon layer, combined with the preponderances of electronic structure optimization of NiO@C, the prepared NiO@C/CC shows great electrocatalytic stability and activity, and can be utilized for the electrooxidation of alcohols, containing ethylene glycol, ethanol, and methanol.

Up to now, Pt nanoparticles or nanoclusters which are supported on conductive carbon substrates have been thought to be the candidates with great promise for MOR catalysts because of the great catalytic activity. But most MOR catalysts have a common challenge: carbonic oxide (CO) poisoning. The dissociation of methanol molecules will induce CO to adsorb to the catalyst surface and passivate the active site, resulting in the degradation of electrocatalytic performance. In order to solve this problem, previous work mainly concentrated on the design and synthesis of surface modified platinum-based catalysts, such as the Clavilier approach to promote the oxidation of CO_{ad}, but the conditions of the synthesis are complex and harsh. Currently, increasing efforts are aimed at designing platinum nanoparticles (NP, approximately 3 nm in size) and nanoclusters (NC, typically ranging from 1 to 3 nm in size) with ultra-fine sizes to maximize the use of platinum and enhance its

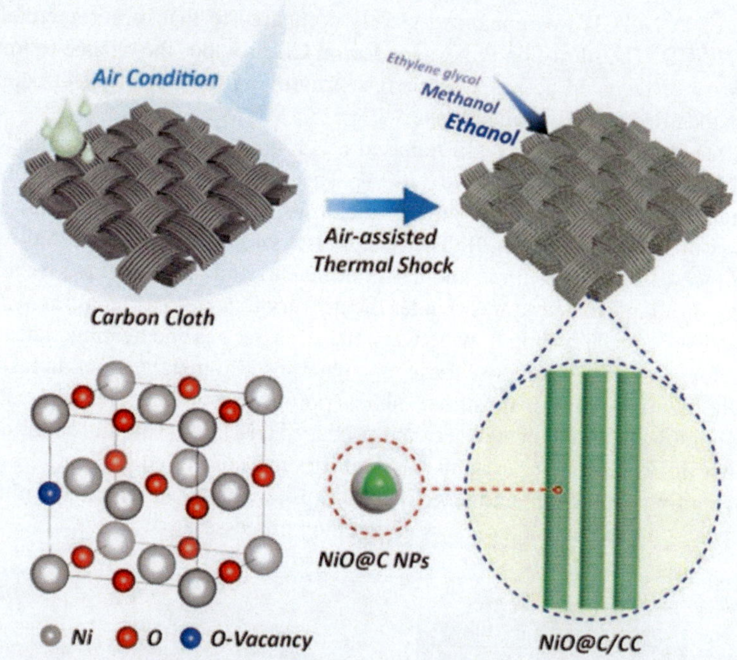

Fig. 2.15 Synthesis path of NiO@C/C. The highly dispersed NiO@C NPs are manufactured in situ on a carbon cloth substrate by air assisted thermal shock technology (ATS-TECH)

electrocatalytic activity in DMFC. While simultaneously shrinking platinum particles and reducing their load seems to be a feasible strategy, conventional wet impregnation and subsequent reduction treatments are not sufficient to simultaneously increase electrochemical activity and reduce costs. So, it is very important to develop an effective synthesis method to obtain ultra-fine platinum nanoparticles and carbon nanoparticles.

Since the nucleation, growth, and aggregation of nanoparticles are the processes of time-dependent diffusion, controlling the sintering time and cooling speed in the synthesis process is critical to obtain the desired particle size and morphology, thus improving the electrochemical activity. In order to achieve this goal, the HTS method based on joule heating has successfully synthesized high-entropy and single-element alloy nanoparticles rapidly (55 ms) at ultra high temperatures (approx. 2000 K) using conductive carbon nanofibers as the substrate of joule heating. However, synthesizing nanoparticles loaded on carbon black by joule heating is challenging because it requests a continuous conductive film to apply an electric current effectively. In 2020, Liangbing Hu 's team reported the synthesis of ultra-fine Pt NC loaded on carbon black (CB) powder substrate at relatively low temperatures using HTS method, with attractive resistance to CO toxicity and catalytic activity to MOR in DMFC [108]. The obtained Pt NC/CB was used to catalyze MOR and showed a low initial potential of 0.4 V and a peak current density of 10.6 mA/cm^2, showing that the catalytic activity of Pt NC/CB was enhanced greatly compared to that of commercial Pt/C and Pt NP/CB. The ultra-fine Pt NCs loaded on CB provides the chance to improve the density of revealed active sites, thereby improving the overall electrocatalytic performance of fuel cell applications.

In 2021, Liangbing Hu's team reported a HTS reactor for synthesizing loaded nanoparticles at temperatures up to 3200 K. This approach can not only be applied to high temperature synthesis of powder materials, but also can achieve successive synthesis of nanoparticles [109]. The reactor is a simple design consisting of two face-to-face carbon paper films about 1–3 mm apart and can be heated by joule to produce ultra high temperatures. Under the action of carrier gas and gravity, metal salt precursors on the carbon flow between these carrier gas and heating plates. High temperatures rapidly decompose these raw materials into metal nanoparticles which are anchored to the carbon substrate, such as porous carbon black particles. By this means, homogeneous nanocatalyst can be produced on a large scale and continuously, as shown in Fig. 2.16. The reactor has uniform and controlled temperature distribution, continuous heating and large reaction space which can realize the ultra-fast decomposition of various precursors in common processes.

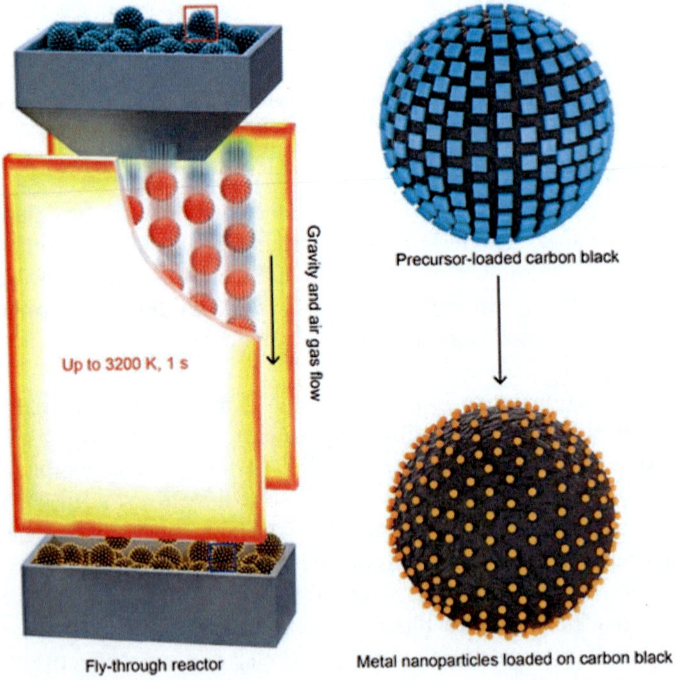

Fig. 2.16 Schematic diagram of a high temperature impact reactor in which the raw material (precursor carbon black powder) flows fast downward between two joule-heated carbon plates under the action of carrier gas and gravity. The precursor salt is thermally decomposed in situ into metal nanoparticles uniformly fixed on carbon black

As a demonstration of definition, the team synthesized Pt nanoparticles anchored to carbon black at 1400 K utilizing this successive fly-through high-temperature reactor. The generated Pt nanoparticles (particle size about 4 nm) were used as anodic catalysts for methanol oxidation, and outstanding electrocatalytic stability and activity were observed. This work provides a fast and common strategy for the preparation of monodisperse and size-controlled nanocatalysts for direct methanol fuel cells and other technologies of renewable energy storage.

In addition to basic direct alcohol fuel cells, direct hydrogen peroxide fuel cells (DPPFC) use hydrogen peroxide (H_2O_2) as cheap fuel for the anode and oxidizer for the cathode, which greatly reduces the expense and improves the feasibility of DPPFC [110]. Hydrogen peroxide has a faster two-electron transfer kinetics than substitute fuels and has no by-products or toxic intermediates during electrooxidation. In addition, the production of oxygen by electrooxidation at the anode is beneficial for anaerobic operating environments, such as underwater power sources. As a result, the research of high-performance materials for electrooxidation of hydrogen peroxide has drawn great attention in the electrochemical catalysis field.

Recently, a variety of types of precious metal catalysts have been investigated for efficient electrooxidation of H_2O_2, for example, gold (Au), silver (Ag), platinum (Pt), palladium (Pd), and their alloys [111–114]. But the high price and inherent scarcity of these precious metals restrict their widespread utilization. As low-cost substitutes, some transition metals (e.g., nickel (Ni) and cobalt (Co)), have been explored as catalysts for H_2O_2 electrooxidation. These metals in the undeveloped state relatively low catalytic properties limits their application in electrocatalysis field, nanostructure can provide higher specific surface area and reveal more active sites for efficient electrocatalysis, this has been demonstrated in recent studies of nanoparticles, nanosheets, nanorods, nanoribbons, and nanowires. However, the presence of adhesives in the electrode nanostructures decreases the electrical conductivity and blocks some catalytic active sites, resulting in bad performance. The adhesive gradually degrades during the electrocatalytic reaction, making the active material separate from the electrode. Hence, designing a nanostructure with a high electrically active surface area on a highly conductive substrate without using an inactive adhesive is an ideal choice for efficient H_2O_2 electrooxidation.

Nanoparticle-bearing carbon substrates, such as graphene and carbon nanotubes, have been reported widely for applications of high-performance energy. Professor Yiju Li/Yanan Chen/Liangbing Hu transformed Ni particles with particle size of about 2 μm into nanoparticles with particle size of about 75 nm in an independent RGO membrane by HTS treatment [115]. After treatment, the nickel nanoparticles are coated in several layers of carbon and uniformly anchored to the surface of the RGO nanosheet (nano-Ni@C/RGO). The nickel nanoparticles with thin carbon coated have highly exposed catalytic sites on the carbon shell, which improves the charge transfer ability and significantly improves the electrooxidation performance of hydrogen peroxide. A good carbon coating protects nickel nanoparticles from unstable bubbles during oxygen generation. The strong bonding between RGO nanosheets and carbon-coated Ni nanoparticles results in great catalytic stability and thus better electrocatalytic performance than the initial nickel particle/RGO composite electrode (Micro-Ni/RGO).

Carbon-coated nickel nanoparticles were formed in situ by joule heating method with ultra-rapid processing at 2370 K for 60 ms, as shown in Fig. 2.17. The resulting nickel nanoparticles do not agglomerate uniformly on the surfaces between RGO nanosheets due to the "trap effect" of the RGO defect, i.e., the heteronucleation position. Conceptually, when heating the composite film above the melting temperature of about 1730 K, the nickel particles melt. When the composite film is cooled abruptly by removing the applied current, the molten nickel will cool into nanoparticles and self-assemble into nanoclusters on the surface of RGO within 0.1 s without aggregation at the vacancy defects and residual oxygen groups. At the same time, due to the catalytic action of nickel, a thin layer of carbon is generated on the surface of nickel nanoparticles during the process of joule heat treatment. The adsorption energy decreases with the increase of the number of carbon atoms, which proves the rationality of enhanced carbon encapsulation. The results of simulation support the formation of carbon layer coated with nickel nanoparticles on RGO nanosheets surface.

The electrochemical performance of the prepared DPPFC anode material electrode was evaluated. Cyclic voltammetry (CV) was used to estimate the specific surface area (ESSA) of the prepared electrode. Nano-Ni@C/RGO, micro-Ni/RGO, and pure RGO electrodes were tested in KOH solution of 4 mol/L. The results showed that there was no REDOX reaction on pure RGO electrode surface. The ESSA values of nano-Ni@C/RGO and micro-Ni/RGO electrodes are 1245.14 cm^2/cm^2 and 29.18 cm^2/cm^2, respectively, and the ESSA values of the electrodes are significantly increased after ultra-fast joule heating treatment. The surface-active centers of thin carbon-coated nickel nanoparticles are highly utilized due to the extremely small size of the nanoparticles. H_2O_2 may react with OH^- on the thin carbon layer surface through electron transfer due to the regulation of Ni nanoparticles in the interior. In the process of H_2O_2 electrooxidation, the current density of nano-Ni@C/RGO electrode increases with the increase of KOH and H_2O_2 concentration. However, higher H_2O_2 concentration (>0.90 mol/L) and KOH concentration (>4 mol/L) did not help to improve the electrocatalytic performance of H_2O_2 electrooxidation reaction. In the test, 0.9 mol/L H_2O_2 and 4 mol/L KOH were used as the best test solutions to test the H_2O_2 electrooxidation performance of nano-Ni@C/RGO, micro-Ni/RGO, and RGO electrodes. The initial oxidation potential of nano-Ni@C/RGO electrode for H_2O_2 electrooxidation is about −0.22 V (vs Ag/AgCl), and the micro-Ni/RGO electrode is −60 mV. The thin carbon layer can reduce the initial potential, and the lower initial oxidation potential shows that the reaction sites on nano-Ni@C/RGO electrode are much more active than those on micro-Ni/RGO electrode. In addition, the oxidation current density of nano-Ni@C/RGO electrode is 602 mA/cm^2 at 0.2 V, which is 143 times higher than that of micro-Ni/RGO electrode at 4.2 mA/cm^2. Therefore, nano-Ni@C/RGO electrode presents excellent electrocatalytic performance for H_2O_2 electrooxidation in terms of initial potential and oxidation current density. These properties are due to the small size of the nanoparticles, the increased use of surface-active sites, the slow kinetics of accelerated electrooxidation by the carbon

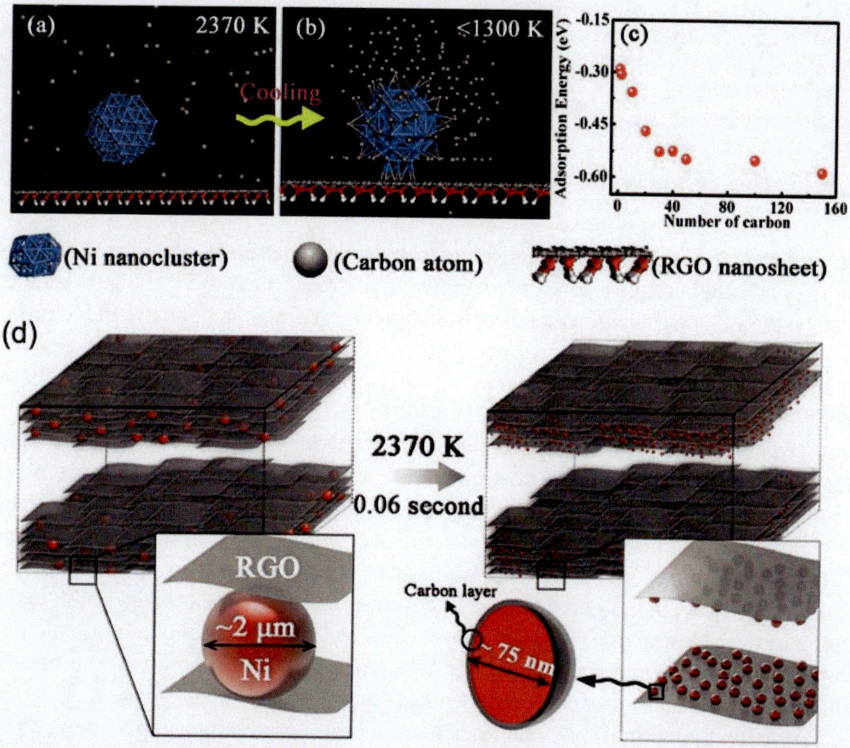

Fig. 2.17 Simulation of carbon layer formation on Ni nanoparticles **a** Ni particles melt at a high temperature of about 2370 K; **b** after cooling, nanoparticles are self-assembled because of defects on RGO nanosheets. Meanwhile, carbon atoms are coated on the surface of nickel nanoclusters during the process of cooling; **c** the adsorption energy reduces with the addition of the carbon adsorption number, which proves that the enhanced carbon encapsulation is reasonable. The simulation results support the formation of carbon layer coated with Ni nanoparticles on the surface of RGO nanosheets; **d** schematic diagram of ultra-fast synthesis of Ni nanoparticles encapsulated in carbon layers on RGO nanosheets. When Ni particles loaded in RGO matrix are heated to ultra high temperature of about 2370 K for about 0.06 s, they reassemble into thin and uniform carbon-coated nanoparticles with an average diameter of 75 nm

shell, and the electrical conductivity of the substrate of RGO nanosheets. Because of the thin carbon layer coating that immobilized Ni nanoparticles on the RGO surface, the nano-Ni@C/RGO material had great stability and the electrooxidation current density decreased slightly after 500 cycles, in 0.9 mol/L H_2O_2 and 4 mol/L KOH solutions at different potentials, the oxidation current density of H_2O_2 remained stable in the test period of 1000 s, showing excellent electrochemical and structural stability. The stable current density further proves that nano-Ni@C/RGO electrode has outstanding cyclic stability and is a promising catalyst for H_2O_2 electrooxidation applications. The newly developed nano-Ni@C/RGO film with high electroactive specific surface area and carbon encapsulation is a promising DPPFC anode material with an economical and scalable process of joule heating. It is expected that the new heating approach demonstrated can be used to prepare other kinds of nanoparticles, with wide applications in the fields of energy storage and conversion.

2.5 Other Catalysts

Many applications such as fuel cells, electrolytic cells, batteries, organic electrosynthesis, and others require catalysts to catalyze the reaction at the cathode or anode. Catalysts based on transition metals or even precious metals can be costly and harmful to the environment if the waste catalysts are not handled properly. The electrochemical cell can accumulate reactants, carbon electrodes, and/or by-products of electrolyte decomposition over long periods of time, blocking the active electrode surface and catalyst, and the by-products can inactivate the surface of electrode and degrade its performance. Due to limited life, a large number of electrochemical devices have been abandoned. Existing recovery methods, for example, fire extraction, hydrometallurgical extraction and fire wet extraction followed by leaching, purification, separation, and resynthesis, extract specific metal components only by spoiling the main metal components. In contrast, less developed regeneration technologies (non-destructive recovery approaches) such as chemical coatings, calcination, and hydrothermal treatment offer better prospects for practice in the future, allowing direct reuse of catalytic electrodes.

Professor Liangbing Hu's team designed a high temperature impact device for high temperature pulse annealing to regenerate catalytic electrode [9]. Figure 2.18 shows the regeneration process. The device applies an electrical pulse to the carbon heating element, rapidly raising the temperature to 1700 K and cooling it to room temperature in milliseconds. High temperature ensures the complete decomposition of the by-products, while rapid annealing keeps the original physical and chemical properties, so as to maintain the performance of the catalyst. The team used a lithium-air battery as a model system. The used Ru-loaded carbon electrode can be reused 10 times directly after each regeneration. The life of the lithium-air battery is extended from about 200 h to about 2,000 h, with considerable overpotential and capacity after each cycle. This research paves the way for the highly sustainable operation of electrochemical devices.

In addition to the above-mentioned nanocatalysts, which can be used for catalytic electrolysis of water and battery reactions, Liangbing Hu's team also prepared high-entropy alloy nanoparticles (HEA-NPs) by HTS, which can be utilized in industrial processes [116–118]. Among them, five-element PtPdRhRuCe HEA-NPs can be used as advanced catalyst for ammonia oxidation, which is the critical process of industrial synthesis of nitric acid. NH_3 is an ideal carrier of H_2. Under the guidance of the calculation method, the large composition space between active metals (Ru, Co, Rh, Pd, Ni, Ir, Fe, Cr, Cu, and M) is explored to catalyze the decomposition of NH_3. The high temperature impact synthesis of Ru-4 MEA NPs can achieve 100% NH_3 decomposition at about 470 °C. However, Ru samples with similar Ru load show only 30% of the conversion rate. The five-component CoMoFeNiCu HEA nanoparticles

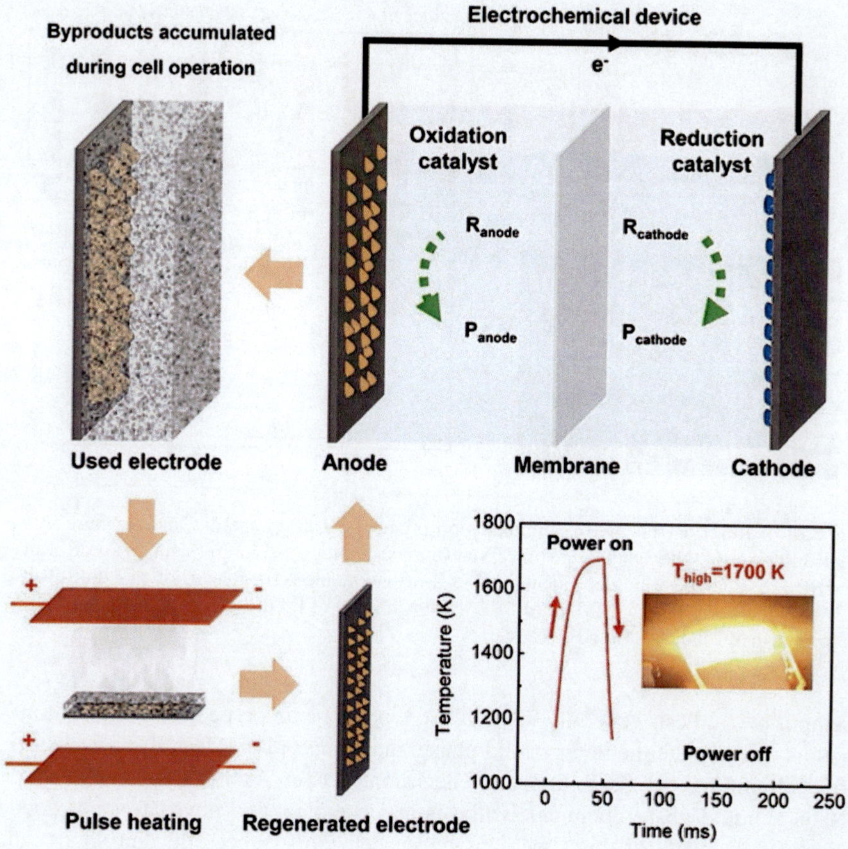

Fig. 2.18 Schematic diagram of regeneration process **a** During the long operation of converting reactants (R) to products (P), the electrochemical cell is subjected to by-product accumulation and contamination at the electrode, during which the catalyst (yellow and blue particles) is inactivated; **b** the regeneration unit consists of two pieces of carbon paper, utilized as a joule heating element, with the used electrodes clamped between them. After regeneration, the electrode surface was exposed again, the catalyst was recovered and reused directly; **c** carbon heaters with rapid heating and cooling speeds (about 10^4 K/s) are appropriate for catalytic electrode regeneration

synthesized by HTS showed significantly enhanced catalytic activity and stability of ammonia decomposition. For example, Fig. 2.19 shows the catalytic performance of 5-member HEA-NPs (PtPdRhRuCe) for ammonia oxidation.

2.6 Summary

The emergence of nanomaterials has led to advances in science and technology, many low-cost and scalable methods for preparing high-quality nanoparticles have been reported, and plenty of types of nanoparticles are available commercially. These

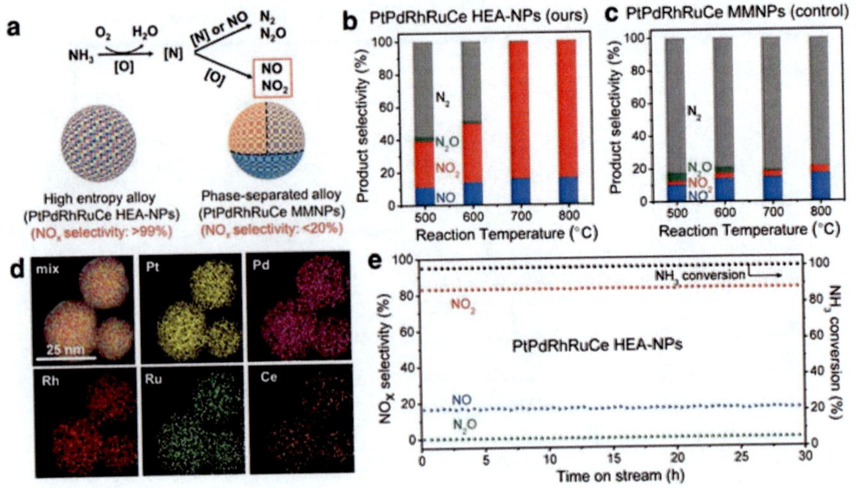

Fig. 2.19 **a** Reaction scheme of ammoxidation process and the properties and structure of the PtPdRhRuCe HEA-NPs synthesized by CTS and the wet-impregnated control sample (PtPdRhRuCe MMNPs); **b**, **c** temperature-dependent product distribution and NH_3 conversion of PtPdRhRuCe HEA-NP and PtPdRhRuCe MMNP; **d** STEM element map of PtPdRhRuCe HEA-NPs; **e** STEM element map of PtPdRhRuCe HEA-NPs

methods of synthesis generally fall into two types. The first type is the "top–bottom" method, usually the method of solid phase, including pulsed laser ablation deposition, ball milling, explosion line, etc. The second type is "bottom-top" methods, for instance, liquid phase chemical synthesis and chemical and physical vapor deposition. While many of these methods have been fairly successful, but significant challenges remain. Traditional synthesis methods tend to synthesize materials with thermodynamic advantages in a relatively mild environment, but it is still difficult to achieve super-fast synthesis of metastable materials under harsh kinetic conditions. For non-noble metals with high activity, nanomaterials synthesized by traditional methods are susceptible to agglomeration and surface oxidation, which significantly limit the function of nanoparticles. Therefore, there is an urgent need for a low-cost, scalable, and fast method to produce uniformly distributed nanoparticles, particularly high-energy nanoparticles without oxidation and agglomeration damage. HTS has great potential and flexible characteristics in application, which can provide rich dynamic conditions for metastable nanomaterials synthesis through fast heating and annealing, and achieve accurate control of the composition and structure of the final product. It has a huge industrial prospect of low cost and high energy efficiency. The new HTS technology has the following advantages: (1) It can provide an extreme environment to achieve ultra-fast synthesis of metastable materials, and reduce the surface oxidation and agglomeration of synthesized nanomaterials; (2) different from the traditional method of wet chemical, this technique can break through the restriction of immiscibility between different metal elements. Rapid heating and quenching at the level of millisecond make it possible to get the maximum configuration entropy of high-entropy alloy (HEA) nanoparticles; (3) the approach is universal. The peak temperature of this method can reach 2000–3000 K, almost as high as the decomposition temperature of most metal salts, which ensures the uniform mixing of many kinds elements in the nanoparticles obtained; (4) this approach can precisely control the crystal structure, composition, grain size and element distribution of nanoalloy by adjusting the temperature, heating time, cooling speed, solution concentration, and defect distribution on carbon carrier; (5) the technology of electric pulse heating saves energy, improves the output efficiency, and shorts the production cycle. It has the potential to achieve large-scale and high-throughput industrial production of nanomaterials. It also opens up a new way for the development of novel polymetallic nanomaterials with broad application prospects. Super-fast synthetic strategy, in short, in the design and synthesis of new nanomaterials have great potential and application prospect, the future of nanostructures super-fast synthetic research may focus on better control of particle size and alloy composition on the synthetic technology of in situ, pattern and super-fast will meet efficient energy storage of high-throughput synthesis and catalytic space.

References

1. Liu L, Chen JF, Yang XF et al (2019) Tailoring pore structure and OER performance of Co_3O_4 mesoporous nanosheets by heat-treatment temperature. Modern Chem Indus 12:66–71 (in Chinese)
2. Choi Y, Ahn T Y, Kim J et al (2020) OER/ORR properties of massively synthesized NiMoS with solvothermal method. ECS Meeting Abstracts, Volume 2020-01, I02: Hydrogen or Oxygen Evolution Catalysis for Water Electrolysis 6
3. Sun T, Liu E, Liang X et al (2015) Enhanced hydrogen evolution from water splitting using Fe-Ni codoped and Ag deposited anatase TiO_2 synthesized by solvothermal method. Appl Surf Sci 347(15):696–705
4. Cong J, Cui D, Jay DS et al (2019) In-situ hydrothermal fabrication of CdS/g-C_3N_4 nanocomposites for enhanced photocatalytic water splitting. Mater Lett 240:128–131
5. Jiang R, Da Y, Han X et al (2021) Ultrafast synthesis for functional nanomaterials. Cell Rep Phys Sci 2(1):100302
6. Kitchen HJ, Vallance SR, Kennedy JL et al (2014) Modern microwave methods in solid-state inorganic materials chemistry: from fundamentals to manufacturing. Chem Rev 114(2):1170–1206
7. Xiao J, Liu P, Wang C et al (2017) External field-assisted laser ablation in liquid: an efficient strategy for nanocrystal synthesis and nanostructure assembly. Progress Mat Sci 87:140–220
8. Zhu Y, Choi SH, Fan X et al (2017) Recent progress on spray pyrolysis for high performance electrode materials in lithium and sodium rechargeable batteries. Adv Energy Mater 7:1–4
9. Dong Q, Li T, Yao Y et al (2020) A general method for regenerating catalytic electrodes. Joule 4(11):2374–2386
10. Kevin MR (2021) The role of hydrogen in our energy future. Climate Energy 37(10):26–32
11. Ankita R, Pichiah S, Min J (2021) Recent progress on visible active nanostructured energy materials for water split generated hydrogen. J Nanostruct Chem 11(1):69–92
12. Lv F, Zhang W, Yang W et al (2020) Ir-based alloy nanoflowers with optimized hydrogen binding energy as bifunctional electrocatalysts for overall water splitting. Small Methods 4(6):1–7

13. Meena N, Vijayalakshmi G, Dinesh K (2021) Metal organic frameworks as electrocatalysts: Hydrogen evolution reactions and overall water splitting. Int J Hydrogen Energy 46(17):10216–10238
14. Subhasis S, Saikat B, Naresh CM et al (2021) An account of the strategies to enhance the water splitting efficiency of noble-metal-free electrocatalysts. J Energy Chem 59:160–190
15. Li Z, Hu M, Wang P et al (202) Heterojunction catalyst in electrocatalytic water splitting. Coordination Chem Rev, 439
16. Xu B, Zhang Y, Pi Y et al (2021) Research progress of nickel-based metal-organic frameworks and their derivatives for oxygen evolution catalysis. Acta Phys Chim Sin 37(7):2009074
17. Su J, Yang Y, Xia G et al (2017) Ruthenium-cobalt nanoalloys encapsulated in nitrogen-doped graphene as active electrocatalysts for producing hydrogen in alkaline media. Nat Commun 8(1):1–10
18. Cao W, Tang Y, Cui Y et al (2020) Energy transfer in metal-organic frameworks and its applications. Small Struct 1(3):2000019
19. Liang K, Yan Y, Guo L et al (2017) Strained $W(Se_xS_{1-x})_{(2)}$ nanoporous films for highly efficient hydrogen evolution. ACS Energy Lett 2(6):1315–1320
20. Zhao J, Chen X, Chen B et al (2019) Accurate control of core-shell upconversion nanoparticles through anisotropic strain engineering. Adv Func Mater 29(44):1903295
21. Jiao W, Chen C, You W et al (2020) Tuning strain effect and surface composition in PdAu hollow nanospheres as highly efficient ORR electrocatalysts and SERS substrates. Appl Catalysis B-Environ 262:118298
22. You B, Tang MT, Tsai C et al (2019) Enhancing electrocatalytic water splitting by strain engineering. Adv Mater 31(17):1807001
23. Luo M, Zhao Z, Zhang Y et al (2019) PdMo bimetallene for oxygen reduction catalysis. Nature 574:7776
24. Alinezhad A, Gloag L, Benedetti TM et al (2019) Direct growth of highly strained Pt islands on branched Ni nanoparticles for improved hydrogen evolution reaction activity. J Am Chem Soc 141(41):16202–16207
25. Xia Z, Guo S (2019) Strain engineering of metal-based nanomaterials for energy electrocatalysis. Chem Soc Rev 48(12):3265–3278
26. Strasser P, Koh S, Anniyev T et al (2010) Lattice-strain control of the activity in dealloyed core-shell fuel cell catalysts. Natural Chem 2(6):454–460
27. Oezaslan M, Hasche F, Strasser P et al (2013) Pt-based core-shell catalyst architectures for oxygen fuel cell electrodes. J Phys Chem Lett 4(19):3273–3291
28. Guo S, Zhang S, Su D et al (2013) Seed-mediated synthesis of core/shell FePtM/FePt (M = Pd, Au) nanowires and their electrocatalysis for oxygen reduction reaction. J Am Chem Soc 135(37):13879–13884
29. Li J, Yin H, Li X et al (2017) Surface evolution of a Pt-Pd-Au electrocatalyst for stable oxygen reduction. Nat Energy 2(8):17111
30. Saleem F, Zhang Z, Cui X et al (2019) Elemental segregation in multimetallic core-shell nanoplates. J Am Chem Soc 141(37):14496–14500
31. Mariano RG, McKelvey K, White HS et al (2017) Selective increase in CO_2 electroreduction activity at grain-boundary surface terminations. Science 358(6367):1187–1191
32. Liu S, Hu Z, Wu Y et al (2020) Dislocation-strained IrNi alloy nanoparticles driven by thermal shock for the hydrogen evolution reaction. Adv Mater 32(48):2006034
33. Wang H, Tsai C, Kong D et al (2015) Transition-metal doped edge sites in vertically aligned MoS_2 catalysts for enhanced hydrogen evolution. Nano Res 8(2):566–575
34. Xie J, Zhang H, Li S et al (2013) Defect-rich MoS_2 Ultrathin nanosheets with additional active edge sites for enhanced electrocatalytic hydrogen evolution. Adv Mater 25(40):5807
35. Li J, Wang Y, Liu C et al (2016) Coupled molybdenum carbide and reduced graphene oxide electrocatalysts for efficient hydrogen evolution. Nat Commun 7:11204
36. Li Y, Wang H, Xie L et al (2011) MoS_2 nanoparticles grown on graphene: an advanced catalyst for the hydrogen evolution reaction. J Am Chem Soc 133(19):7296–7299

37. Deng J, Li H, Xiao J et al (2015) Triggering the electrocatalytic hydrogen evolution activity of the inert two-dimensional MoS_2 surface via single-atom metal doping. Energy Environ Sci 8(5):1594–1601
38. Merki D, Vrubel H, Rovelli L et al (2012) Fe Co, and Ni ions promote the catalytic activity of amorphous molybdenum sulfide films for hydrogen evolution. Chem Sci 3(8):2515–2525
39. Merki D, Fierro S, Vrubel H et al (2011) Amorphous molybdenum sulfide films as catalysts for electrochemical hydrogen production in water. Chem Sci 2(7):1262–1267
40. Cui X, Ren P, Deng D et al (2016) Single layer graphene encapsulating non-precious metals as high-performance electrocatalysts for water oxidation. Energy Environ Sci 9(1):123–129
41. Wu H, Xia B, Yu L et al (2015) Porous molybdenum carbide nano-octahedrons synthesized via confined carburization in metal-organic frameworks for efficient hydrogen production. Nat Commun 6:6512
42. Kong D, Wang H, Cha JJ et al (2013) Synthesis of MoS_2 and $MoSe_2$ films with vertically aligned layers. Nano Lett 13(3):1341–1347
43. Lukowski MA, Daniel AS, English CR et al (2014) Highly active hydrogen evolution catalysis from metallic WS_2 nanosheets. Energy Environ Sci 7(8):2608–2613
44. Garcia ET, Cha D, Ou Y et al (2013) Tungsten carbide nanoparticles as efficient cocatalysts for photocatalytic overall water splitting. Chemsuschem 6(1):168–181
45. Wang H, Kong D, Johanes P, et al (2013) $MoSe_2$ and WSe_2 nanofilms with vertically aligned molecular layers on curved and rough surfaces. Nano Lett 2103, 13(7):3426–3433
46. Liu W, Hu E, Jiang H et al (2016) A highly active and stable hydrogen evolution catalyst based on pyrite-structured cobalt phosphosulfide. Nat Commun 7:10771
47. Wang H, Lee HW, Deng Y et al (2015) Bifunctional non-noble metal oxide nanoparticle electrocatalysts through lithium-induced conversion for overall water splitting. Nat Commun 6:7261
48. Feng L, Yu G, Wu Y et al (2015) High-index faceted Ni_3S_2 nanosheet arrays as highly active and ultrastable electrocatalysts for water splitting. J Am Chem Soc 137(44):14023–14026
49. Pu Z, Liu Q, Tang C et al (2014) Ni_2P nanoparticle films supported on a Ti plate as an efficient hydrogen evolution cathode. Nanoscale 6(19):11031–11034
50. Chen W, Sasaki K, Ma C et al (2012) Hydrogen-evolution catalysts based on non-noble metal nickel-molybdenum nitride nanosheets. Angewandte Chemie-Inrernational Edition 51(25):6131–6135
51. McKone JR, Sadtler BF, Werlang CA et al (2013) Ni-Mo nanopowders for efficient electrochemical hydrogen evolution. ACS Catal 3(2):166–169
52. Caban AM, Kaiser NS, English CR et al (2014) Ionization of high-density deep donor defect states explains the low photovoltage of iron pyrite single crystals. J Am Chem Soc 136(49):17163–17179
53. Anna D, Rachel C, Landon O et al (2015) Ultrafine iron pyrite (FeS_2) nanocrystals improve sodium-sulfur and lithium-sulfur conversion reactions for efficient batteries. ACS Nano 9(11):11156–11165
54. Susac D, Zhu L, Teo M et al (2007) Characterization of FeS_2-based thin films as model catalysts for the oxygen reduction reaction. J Phys Chem C 111(50):18715–18723
55. Kong D, Cha J, Wang H et al (2013) First-row transition metal dichalcogenide catalysts for hydrogen evolution reaction. Energy Environ Sci 6:3553–3558
56. Chen Y, Xu S, Li Y et al (2017) FeS_2 nanoparticles embedded in reduced graphene oxide toward robust, high-performance electrocatalysts. Adv Energy Mater 7(19):1700482
57. Suen N, Hung S, Quan Q et al (2017) Electrocatalysis for the oxygen evolution reaction: recent development and future perspectives. Chem Soc Rev 46:337–365
58. Cheng F, Chen J (2012) Metal-air batteries: from oxygen reduction electrochemistry to cathode catalysts. Chem Soc Rev 41:2172–2192
59. Hannah O, Surya D, Xu H et al (2016) Transition metal (Fe Co, Ni, and Mn) oxides for oxygen reduction and evolution bifunctional catalysts in alkaline media. Nano Today 11(5):601–625
60. Gong M, Dai H (2015) A mini review of NiFe-based materials as highly active oxygen evolution reaction electrocatalysts. Nano Res 8:23–39

61. Hong W, Kitta M, Xu Q (2018) Bimetallic MOF-derived FeCo-P/C nanocomposites as efficient catalysts for oxygen evolution reaction. Small Methods 2(12):1800214
62. Yu J, Li Q, Li Y et al (2016) Ternary metal phosphide with triple-layered structure as a low-cost and efficient electrocatalyst for bifunctional water splitting. Adv Func Mater 26(42):7644–7651
63. Zhang T, Zhu Y, Lee J (2018) Unconventional noble metal-free catalysts for oxygen evolution in aqueous systems. J Mat Chem A 6:8147–8158
64. Gan Q, Wu Z, Li X et al (2018) Structure and electrocatalytic reactivity of cobalt phosphosulfide nanomaterials. Top Catal 61:958–964
65. Yan Y, Xia B, Xu Z et al (2014) Recent development of molybdenum sulfides as advanced electrocatalysts for hydrogen evolution reaction. ACS Catal 4(6):1693–1705
66. Bai Z, Heng J, Zhang Q, et al (2018) Rational design of dodecahedral $MnCo_2O_{4.5}$ hollowed-out nanocages as efficient bifunctional electrocatalysts for oxygen reduction and evolution. Adv Energy Mat 8(34):1802390
67. Feng J, Xu H, Dong Y et al (2016) FeOOH/Co/FeOOH hybrid nanotube arrays as high-performance electrocatalysts for the oxygen evolution reaction. Angewandte Chemie-International Edition 128(11):3758–3762
68. Alexis G, Oscar DM, Han B et al (2017) Activating lattice oxygen redox reactions in metal oxides to catalyse oxygen evolution. Nat Chem 9:457–465
69. Shahana C, Rachel C, Landon O, et al (2014) Electrochemical and corrosion stability of nanostructured silicon by graphene coatings: toward high power porous silicon supercapacitors. J Phys Chem C 118(20):10893–10902
70. Ma T, Dai S, Jaroniec M et al (2014) Metal-organic framework derived hybrid Co_3O_4-carbon porous nanowire arrays as reversible oxygen evolution electrodes. J Am Chem Soc 136(39):13925–13931
71. Yang C, Cui M, Li N et al (2019) In situ iron coating on nanocatalysts for efficient and durable oxygen evolution reaction. Nano Energy 63:103855
72. Thomas B, Jack KP, Simon HW et al (2019) High-entropy alloys as a discovery platform for electrocatalysis. Joule 3(3):834–845
73. Paulin B, Jacky R, Pierre B (2015) Multimetallic catalysis based on heterometallic complexes and clusters. Chem Rev 115(1):28–126
74. Xin B, Rosalie KH, Si Z et al (2020) Capturing the active sites of multimetallic (oxy)hydroxides for the oxygen evolution reaction. Energy Environ Sci 13:4225–4237
75. Samuel T, Alfonso D, Eduardo SL et al (2019) Vanadium in biological action: chemical, pharmacological aspects, and metabolic implications in diabetes mellitus. Biol Trace Elem Res 188:68–98
76. Daniel BP, Antonia IM, Elena RA et al (2017) Zirconium phosphate heterostructures as catalyst support in hydrodeoxygenation reactions. Catalysts 7(6):176
77. Matthew WK, Daniel GN (2008) In situ formation of an oxygen-evolving catalyst in neutral water containing phosphate and Co^{2+}. Science 321(5892):1072–1075
78. Yi Z, Ye J, Kikugawa N et al (2010) An orthophosphate semiconductor with photooxidation properties under visible-light irradiation. Nat Mater 9:559–564
79. Qiao H, Wang X, Dong Q et al (2021) A high-entropy phosphate catalyst for oxygen evolution reaction. Nano Energy 86:106029
80. Zhao S, Wang Y, Dong J et al (2016) Ultrathin metal-organic framework nanosheets for electrocatalytic oxygen evolution. Nat Energy 1:16184
81. HyukSu H, Kang MK, Heechae C et al (2018) Parallelized reaction pathway and stronger internal band bending by partial oxidation of metal sulfide-graphene composites: important factors of synergistic oxygen evolution reaction enhancement. ACS Catal 8(5):4091–4102
82. Zhaoa X, Shanga X, Quana Y et al (2017) Electrodeposition-solvothermal access to ternary mixed metal Ni-Co-Fe sulfides for highly efficient electrocatalytic water oxidation in alkaline media. Electrochim Acta 230:151–159
83. Cui Z, Chen H, Zhao M et al (2016) High-performance Pd_3Pb intermetallic catalyst for electrochemical oxygen reduction. Nano Lett 16(4):2560–2566

84. Zhi WS, Kibsgaard J, Dickens CF, et al (2017) Combining theory and experiment in electrocatalysis: Insights into materials design. Nature 355(6321)
85. Du X, Huang J, Zhang J et al (2018) Modulating electronic structures of inorganic nanomaterials for efficient electrocatalytic water splitting. Angewandte Chemie-International Edition 58(14):4484–4502
86. Meenakshi C, Kasala PR, Chinnakonda SG et al (2017) Copper cobalt sulfide nanosheets realizing a promising electrocatalytic oxygen evolution reaction. ACS Catal 7(9):5871–5879
87. Bharath BR, Pranaw K, Pavana SK et al (2019) Accumulation-driven unified spatiotemporal synthesis and structuring of immiscible metallic nanoalloys. Matter 1(6):1606–1617
88. Cui M, Yang C, Li B et al (2020) High-entropy metal sulfide nanoparticles promise high-performance oxygen evolution reaction. Adv Energy Mater 11(3):1–8
89. Wu H, Lu H, Zhang J et al (2020) Thermal shock-activated spontaneous growing of nanosheets for overall water splitting. Nano-Micro Lett 12(1):1–12
90. Xu S, Chen Y, Li Y et al (2017) Universal, in situ transformation of bulky compounds into nanoscale catalysts by high-temperature pulse. Nano Lett 17:5817–5822
91. Chen Y, Xu S, Zhu S et al (2019) Millisecond synthesis of CoS nanoparticles for highly efficient overall water splitting. Nano Res 12(9):2259–2267
92. Tan P, Chen B, Xu H et al (2017) Flexible Zn- and Li-air batteries: recent advances, challenges, and future perspectives. Energy Environ Sci 10:2056–2080
93. Zhou M, Wang HL, Guo S (2016) Towards high-efficiency nanoelectrocatalysts for oxygen reduction through engineering advanced carbon nanomaterials. Chem Soc Rev 45:1273–1307
94. Liu Z, Zhao Z, Peng B et al. Beyond extended surfaces: understanding the oxygen reduction reaction
95. Chung DY, Jun SW, Yoon G et al (2015) Highly durable and active PtFe nanocatalyst for electrochemical oxygen reduction reaction. J Am Chem Soc 137:15478–15485
96. Lai J, Huang B, Tang Y et al (2018) Barrier-free interface electron transfer on PtFe-Fe_2C Janus-like nanoparticles boosts oxygen catalysis. Chem 4:1153–1166
97. Wang Q, Chen FY, Guo LF et al (2019) Nanoalloying effects on the catalytic activity of the formate oxidation reaction over AgPd and AgCuPd aerogels. J Mat Chem A 7:16122–16135
98. Zeng S, Lv B, Qiao J et al (2019) PtFe alloy nanoparticles confined on carbon nanotube networks as air cathodes for flexible and wearable energy devices. ACS Appl Nano Mat 2:7870–7879
99. Fenton JL, Steimle BC, Schaak RE (2018) Tunable intraparticle frameworks for creating complex heterostructured nanoparticle libraries. Science 360:513–517
100. Chen G, Zhao Y, Fu G et al (2014) Interfacial effects in iron-nickel hydroxide-platinum nanoparticles enhance catalytic oxidation. Science 344:495–499
101. Zhang B, Zheng X, Voznyy O et al (2016) Homogeneously dispersed multimetal oxygen-evolving catalysts. Science 352:333–337
102. Batchelor T, Pedersen JK, Winther SH et al (2019) High-entropy alloys as a discovery platform for electrocatalysis. Joule 3(3):834–845
103. Yao Y, Huang Z, Li T et al (2020) High-throughput, combinatorial synthesis of multimetallic nanoclusters. Proc Natl Acad Sci USA 117(12):6316–6322
104. Xie H, Liu Y, Li N et al (2021) High-temperature-pulse synthesis of ultrathin-graphene-coated metal nanoparticles. Nano Energy 80:105536
105. Zhu C, Yang W, Di J et al (2021) CoNi nanoparticles anchored inside carbon nanotube networks by transient heating: low loading and high activity for oxygen reduction and evolution. J Energy Chem 5:63–71
106. Zhu Y, Liu X, Jin S et al (2019) Anionic defect engineering of transition metal oxides for oxygen reduction and evolution reactions. J Mat Chem A 7:5875–5897
107. Liu C, Zhou W, Zhang J et al (2020) Air-assisted transient synthesis of metastable nickel oxide boosting alkaline fuel oxidation reaction. Adv Energy Mater 10(46):1–8
108. Qiao Y, Liu Y, Liu Y et al (2020) Thermal radiation synthesis of ultrafine platinum nanoclusters toward methanol oxidation. Small Methods 4(9):1–7

109. Qiao Y, Chen C, Liu Y et al (2021) Continuous fly-through high-temperature synthesis of nanocatalysts. Nano Lett 21(11):4517–4523
110. Ye K, Guo F, Gao Y et al (2015) Three-dimensional carbon- and binder-free nickel nanowire arrays as a high-performance and low-cost anode for direct hydrogen peroxide fuel cell. J Power Sources 300(30):147–156
111. Li J, Yu Q, Peng T (2005) Electrocatalytic oxidation of hydrogen peroxide and cysteine at a glassy carbon electrode modified with platinum nanoparticle-deposited carbon nanotubes. Nalytical Sci 21(4):377–381
112. Yang F, Cheng K, Wu T et al (2013) Dendritic palladium decorated with gold by potential pulse electrodeposition: Enhanced electrocatalytic activity for H_2O_2 electroreduction and electrooxidation. Electrochim Acta 99(1):54–61
113. Yamazaki S, Siroma Z, Senoh H et al (2008) A fuel cell with selective electrocatalysts using hydrogen peroxide as both an electron acceptor and a fuel. J Power Sources 178(1):20–25
114. Masataka H, Takuro K, Hideaki K (1983) On the electrochemical behavior of H_2O_2 AT Ag in alkaline solution. Electrochim Acta 28(5):727–733
115. Li Y, Chen Y, Nie A et al (2017) In situ, fast, high-temperature synthesis of nickel nanoparticles in reduced graphene oxide matrix. Adv Energy Mater 7(11):1601783
116. Xie P, Yao Y, Huang Z et al (2019) Highly efficient decomposition of ammonia using high-entropy alloy catalysts. Nat Commun 10(1):4011
117. Yao Y, Huang Z, Xie P et al (2018) Carbothermal shock synthesis of high-entropy-alloy nanoparticles. Science 359(6383):1489–1494
118. YaoY, Liu Z, Xie P et al (2020) Computationally aided, entropy-driven synthesis of highly efficient and durable multi-elemental alloy catalysts. Sci Adv 6(11):eaaz0510

Chapter 3
Application of HTS in Rechargeable Battery

Check for updates

Contents

3.1 Introduction

With the industry developing and the climate changing rapidly, devices of energy storage with high performance, environmental protection, low cost, flexibility, and miniature have become much more significant than ever [1]. Among the diversified technologies of energy storage, rechargeable batteries are characterized by flexible power, simple maintenance, and high-energy conversion efficiency and are thought to take place of gasoline as new power source [2]. Fabrication of electrode nanomaterials is a critical element in rechargeable batteries development. And among the rechargeable batteries, lithium-ion batteries (LIBs) are the energy storage devices with great promise due to the advantages of high energy density and long cycle life [3]. Since SONY of Japan commercialized LIBs in the 1990s, the system has been widely applied in portable electronic devices, electro mobile, and large-scale systems of energy storage, and is known as the green energy and leading power

Y. Chen, *High Temperature Shock Technology*, Nanostructure Science and Technology,
https://doi.org/10.1007/978-981-19-8124-1_3

in the twenty-first century [4]. John Goodenough and Stanley Whittingham from the USA and Akira Yoshino from Japan have also been awarded the 2019 Nobel Prize in Chemistry for the achievements on LIBs, making a "fossil-fuel-free society" possible. However, the application of LIBs in the systems of energy storage in large scale is severely restricted due to the limitation of global lithium reserves (only 0.0017 wt.%) and the increasing price of lithium resources caused by uneven distribution [5]. As a result, there is an urgent requirement to develop a new high energy density and low-cost system to replace LIBs to achieve sustainable development in energy storage field. Sodium and potassium of the same main group have similar physical and chemical properties to lithium, and are abundant and diffusely distributed in the earth's crust. And unlike lithium, neither sodium nor potassium can be alloyed with aluminum. Cheap aluminum foil can be utilized as a current collector in Na^+/K^+ batteries, which gives them a great price advantage in terms of production cost [5]. Therefore, we consider sodium/potassium ion batteries as a promising battery technology that can replace LIBs for low-cost and large-scale energy storage utilizations. However, compared with Li^+ ion radius, Na^+/K^+ has a larger ion radius, which will result in a larger volume change of electrode materials in the charging and discharging procedure, a significant decline in kinetic performance and different reaction mechanism, bringing great challenges to the development of Na^+/K^+ batteries. Moreover, developing novel electrode nanomaterials to meet the growing need for LIBs with higher energy density and longer cycle life is still a huge challenge [6]. As is known to all that combining carbon-based materials (such as carbon nanofiber (CNF), reduced graphene oxide (RGO), carbon nanotube (CNT), and porous carbon) with active nanomaterials could significantly improve batteries electrochemical performance [7]. Traditional electrode nanomaterials synthesis methods (e.g., solvothermal, sol–gel, and chemical vapor deposition) may impair nanomaterials properties due to inevitable agglomeration, surface oxidation, and solvent contamination.

HTS is a simple, rapid, and efficient method for manufacturing high-quality and uniformly dispersed nanomaterials, including single metal such as Sn, Ni, and Ag, and semiconductors such as Si, fixed on carbon-based carriers such as CNF and RGO. As a self-supporting electrode in various rechargeable battery systems (e.g., lithium-ion batteries, lithium metal batteries, and Li-CO_2 batteries), it shows great performance, such as long cycle life and high capacity. In addition, carbon-based materials (such as carbonized biomass and RGO or CNT films) and lithium-ion conductive ceramics which are manufactured through the high-temperature superconducting process can also be utilized to the devices of energy storage (such as lithium/sodium ion batteries and lithium solid batteries). The nanomaterials do not need extra adhesives and conductive additives, which helps to increase the rate capacity and energy density of the electrodes. Therefore, the wide application and broad prospect of HTS in energy conversion and storage will offer new chances to develop the next generation of energy-related equipment.

3.2 Preparation of Anode Materials of Rechargeable Battery by HTS

3.2.1 Preparation of Anode Materials of Lithium-Ion Battery by HTS

Lithium-ion batteries (LIBs) are of particular importance for electronic and electric vehicles in the future because of their advantages of high energy density and long cycle life. In order to further improve the energy density of LIBs, scientists are studying a series of electrode materials with high capacity. The volume of the high-capacity composite anode (e.g., Si) changes during lithium and delithium process, resulting in reduced electrochemical performance. Nanostructures can effectively overcome this problem, but are prone to be agglomerated due to the high surface area.

In this section, Professors Yanan Chen/Liangbing Hu [8] used commercial silicon particles (diameter ≈ 2 μm) as precursors to transform silicon microparticles into ultrafine nanoparticles (diameter about 15 nm) embedded in conductive RGO matrix through an in situ synthesis method based on HTS. Because of the RGO nanosheets have high broadband absorption, this method can be applied to a series of electrode materials of battery with high capacity, (e.g., Al and Sn in RGO). In addition, nanoparticles synthesized in situ disperse well in RGO matrix and will not agglomerate, which is beneficial to the utilizations of battery.

As shown in Fig. 3.1, self-supporting RGO films with embedded silicon micron particles (SiMPs) are manufactured through an easy filtration procedure which is accompanied by preheating reduction. The RGO paper which is based on solution can be used as a planar radiant heater, heating it to high temperatures by joule heating and adjusting it by applying the input power. Place the self-supporting RGO-SiMPs film on the top of the high temperature heater. Then the RGO paper below radiatively heated the RGO-SiMPs film for 30 s. Finally, SiMPs are converted into SiNPs (silicon nanoparticles) in situ, and SiNPs are dispersed in highly conductive RGO matrix. In this paper, Sn and Al nanoparticles were synthesized directly in conductive matrix by the same method. These LIB anode particles with high capacity are dispersed well in the conductive RGO, and these negatives can be utilized in high-performance batteries straightway.

Figure 3.2a presents the fast in situ synthesis of RGO-SiNPs by RGO-SiMPs through high temperature treatment with radiation. The RGO paper is used as a radiant heat source while the applied current creates a source of high temperature for nanoparticle thermal reduction and synthesis [9–11]. The RGO-SiMPs films were prepared by vacuum filtration of GO-SiMPs solution and then prereduced at 573 K for 1 h in a tube furnace which is filled with Ar (Fig. 3.2b). Because of RGO nanosheets has high absorption coefficient (similar to the behavior of graphene), RGO-SiMPs films can be heated to high temperature quickly when they are put on the hot RGO paper [12]. The RGO-SiMPs films have a high optical absorptance

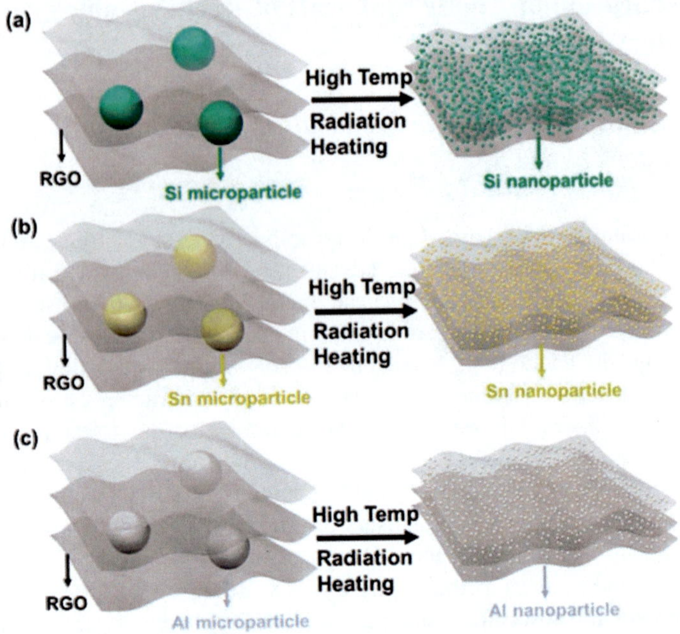

Fig. 3.1 Schematic diagram of a novel nanoparticle (NP) synthesis method for battery utilizations

which can reach 70–86% in the 200–2500 nm range (Fig. 3.2c), indicating that the proposed high-temperature radiation process has excellent heat absorption capacity. Remove the RGO-SiMPs film from the hot RGO paper for a short time and cool it to the room temperature. To quantify the temperature of RGO paper, they used spectrometers. The fiber was put on the top of RGO paper in order to detect and record the RGO paper's emission spectra at different current values. Emission spectra in the wavelength range from 340 to 950 nm (Fig. 3.2d) were used and fitted to the blackbody radiation equation to obtain the corresponding temperature. The spectral radiation intensity increases significantly with the increase of the current, indicating that the temperature is also higher. According to the temperature-power curve, it can be observed that the fitting temperature increases almost linearly with the electrical power going up (Fig. 3.2e).

In order to study the morphological changes of RGO-SiMPs films before and after HTS treatment, SEM is used to measure the morphology of RGO-SIMPs films. Typical RGO-SiMPs films morphology is shown in Fig. 3.3a, from which it can be seen that RGO nanosheets with random aggregation and wrinkling are closely related to irregular SiMPs. The size distribution of the original SiMPs ranges from 0.5 to 2.5 μm, with a mean diameter of 1.5 μm (Fig. 3.3b). The RGO-SiNPs film was formed after HTS treatment of RGO-SiMPs film for 30 s at about 1800 K. Figure 3.3c, d presents SEM images of uniformly distributed SiNPs in RGO. Due to HTS treatment, SiMPs are converted to SiNPs with a small size distribution which are between 10

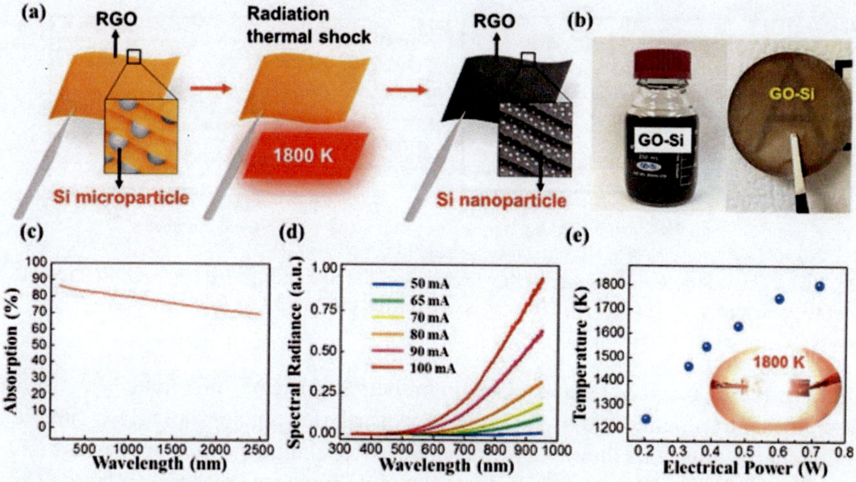

Fig. 3.2 **a** Self-made device for the synthesis of anode nanoparticle composites; **b** digital images of GO-SiMPs solution and prepared GO-SiMPs films; **c** light absorption measurement of RGO-SiMPs film; **d** spectral radiation measurement of RGO paper under different electric power input; **e** fitting temperature as a function of RGO paper input electrical power

and 15 nm (Fig. 3.3e). The obtained RGO-SiNPs films exhibit excellent mechanical strength and flexibility (Fig. 3.3f).

The following transformation mechanism was proposed during HTS of RGO-SiMPs films. Due to thermal radiation, SiMPs melt into liquid Si and are distributed throughout the RGO matrix. Because of the defects on RGO nanosheets, molten

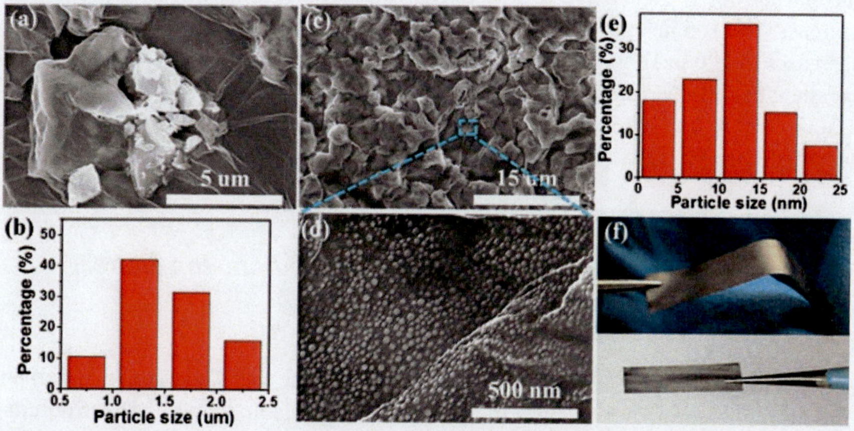

Fig. 3.3 **a** SEM images of RGO-SiMPs films; **b** statistical granularity distribution of SiMPs in RGO network before HTS treatment; **c**, **d** SEM images of RGO-SiNPs film after 30 s HTS treatment at about 1800 K; **e** statistical particle size distribution of SiNPs after HTS **f** Digital image showing flexibility of RGO-SiNPs film after HTS

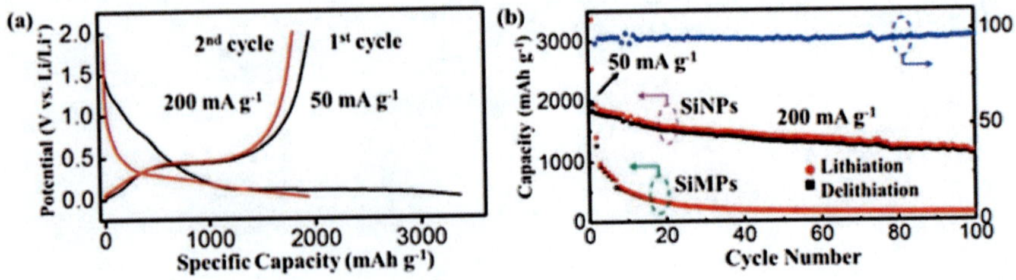

Fig. 3.4 **a** Constant-current discharge/charge curves of the first and second cycles of RGO-SiNPs film; **b** cyclic properties of RGO-SiNPs and RGO-SiMPs films at 200 mA g^{-1}

liquid Si is stuck in RGO. After cooling, molten Si nucleates precipitates and self-assembles into superfine SiNPs and is evenly distributed on the whole RGO surface.

In order to prove that thermal impact nanoparticle composites can be used as high-capacity anode of LIBs, the RGO composite membrane was placed in a button cell and acted as an independent, carbon-free/adhesive-free electrode. The RGO-SiNPs films electrochemical properties were measured by galvanostatic discharge–charge method. Figure 3.4a exhibits the charge–discharge curves of the first two cycles of RGO-SiNPs anode. When the current density is 50 mA g^{-1}, the first discharge platform is 0.1 V and then reduced to 0.01 V gradually. This process corresponds to the insertion of Li into crystalline Si and the generation of amorphous Li_xSi phase [13, 14]. The charging platform of the first cycle is 0.42 V, corresponding to the process of delithium. The subsequent curves of discharge and charge present the characteristic voltage curve of Si, which is in keeping with former researches [15, 16]. The discharge capacity and charge capacity of the first cycle are 3367 mAh g^{-1} and 1957 mAh g^{-1}, respectively, and the first coulomb efficiency is 58%. When the current density is 200 mA g^{-1} (Fig. 3.4b), the capacity of RGO-SiNPs electrode decreases to 1165 mAh g^{-1} after 100 cycles, while the capacity of RGO-SiMPs film decreases to 126 mAh g^{-1} after 100 cycles. It is observed that RGO-SiNPs film has excellent cycling performance, which is improved because the ultrafine SiNPs and RGO matrix reduce strain and alleviate the change of Si volume in the process of Li deembedding [17, 18].

3.2.2 *Preparation of Anode Materials of Sodium-Ion Battery by HTS*

Sodium-ion batteries (SIBs) are widely studied because there is more sodium than lithium in the earth, and because they work in a semblable way to LIBs. Sodium-ion batteries are not only used as a substitute for LIBs in energy storage equipment for large scale, but also compatible with existing equipment used to produce LIBs. Moreover, the sodium-ion battery can use a lower concentration of electrolyte; at the

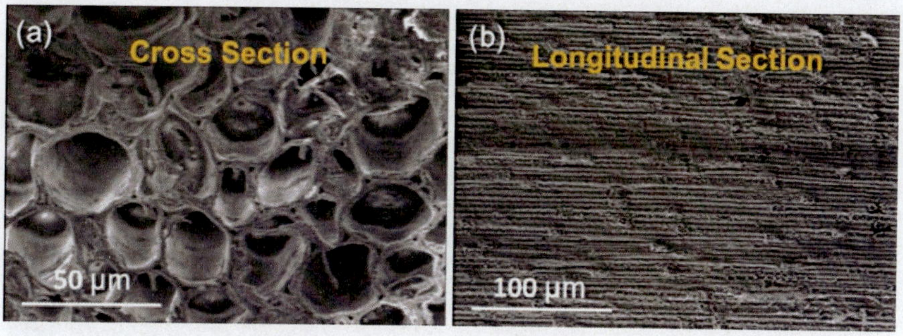

Fig. 3.5 Hierarchy of switchgrass stems **a** SEM image of cross section; **b** the longitudinal section on the stem

same time, the sodium-ion battery does not have to worry about discharge and other problems.

Recently, carbon materials which are biomass derived have drawn much attention because of their environmentally friendly properties and rich resources [19–22]. Switchgrass, a high-yield biomass crop, has been facilitated as the main biomass source for biofuels [23]. Switchgrass is also rich in cellulose, and its stem contains up to 50% cellulose and about 21% crosslinked lignin [24]. The SEM image of the switchgrass cross section in Fig. 3.5a reveals a macroporous structure composed of cell walls to make up an interconnected open frame. SEM observations in the vicinity of the straw sheath shown in Fig. 3.5b highlight the parallel channel array structure in the longitudinal dimension. The rich cellulose with this macrostructure makes switchgrass a desired precursor for producing 3D porous carbon materials.

This section describes how Liangbing Hu's team [25] carbonized switchgrass stems at ultra high temperatures (2050 °C) to produce hard carbon materials. The temperature in this carbonization process is much higher than that of other biomass applied to SIBs [19, 21, 22, 26]. Carbonized switchgrass stems at temperatures in excess of 2000 °C as SIBs anode material. The electrical conductivity of the switchgrass derived carbon increases with the temperature of carbonization increasing, while the surface area decreases. As expected, the carbon anode treated at 2050 °C showed higher first-cycle Coulomb efficiency and rate performance compared with carbon which was prepared at 1000 °C. Besides, the high temperature-induced carbon anode provides a 200 mAh g^{-1}specific capacity in 800 charge/discharge cycles. We have reasons to believe that switchgrass could provide a rich carbon source that could easily produce 3D porous carbon materials which have sodium storage properties. Cut dried switchgrass stems into small sections and wash them with water or ethanol and then dry at 60 °C. Then place the dried grass stems in a tube furnace and carbonized at 1000 °C in an atmosphere of argon for 2 h, named GC-1000. Use the homemade device (Fig. 3.6) to further treat the GC-1000 with joule heating. The GC-1000 rod (3 cm × 1.8 mm) is linked to a copper electrode via a silver paste and suspended on a glass bracket. Both ends of the glass bracket are wired to an outside connected KEITHLEY power supply. The process of joule heating takes place in

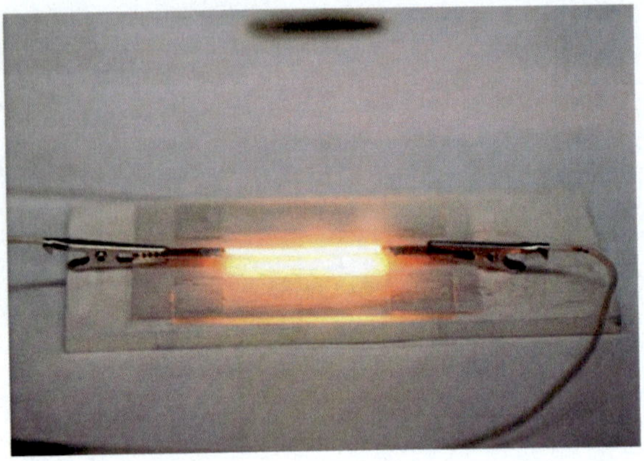

Fig. 3.6 Carbon derived from switchgrass stem under joule heating

a glove box filled with argon, where a DC current is used and the GC-1000 rod is heated at 2050 °C (GC-2050) through joule heating.

It is clear from Fig. 3.7a that the prepared sample has a hollow 3D structure with connections and large pores, which has similarities to the large pore characteristics of grass before carbonization. After carbonization at high temperature, many voids are generated on the carbon wall (Fig. 3.7b). This unique open structure is particularly helpful in promoting ion diffusion and electrolyte penetration.

Fast and efficient HTS technology and excellent sodium-ion storage properties, together with sustainable biomass sources and scalable synthesis methods, make switchgrass derived carbon become an appealing anode material of SIBs.

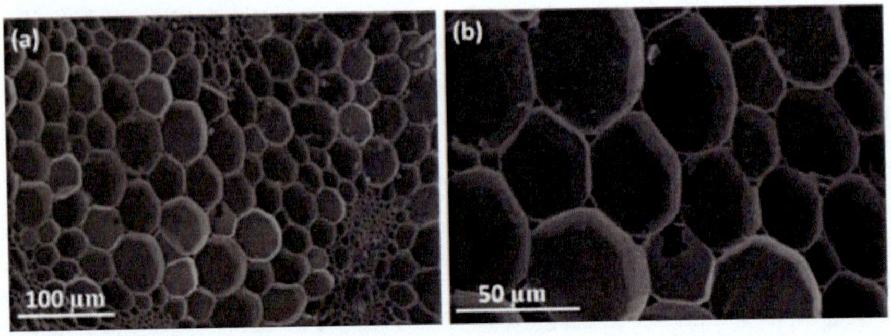

Fig. 3.7 **a** Low magnification SEM image of cross section of hard carbon material; **b** SEM images with higher magnification corresponding to hard carbon materials

3.3 Preparation of Current Collector of Rechargeable Battery by HTS

3.3.1 Preparation of Current Collector of Lithium-Ion Battery by HTS

The demand for highly conductive and lightweight films is everywhere for electronic and energy equipment. For instance, for the current collector utilized in LIBs, the negative electrode is copper (Cu) foil, and the positive electrode is aluminum (Al) foil, accounting for 15–50% of the overall weight of the device [27]. Current collector account for 13% of cost and 6% of weight in batteries (e.g., NiMH), but they act a key role in the performance of electrochemical (EC), especially EC stability and interfacial stability of batteries. Current collector used in water-based batteries (titanium (Ti), stainless steel (SS), nickel (Ni) foams, carbon cloth, and nickel-plated SS) are costly, bulky and prone to be corroded after prolonged utilization [28, 29]. RGO nanosheets are very charming because of their big transverse size after processing (reach several hundred microns), resulting in much lower permeability thresholds and fewer connections in continuous films, resulting in high conductivity [30, 31]. Achieving high purity by reducing graphene oxide close to original graphene is a huge challenge, because defects and functional groups in partially reduced RGO will block the transport of the carriers largely. A variety of reduction approaches have been put forward, containing chemical reduction thermal reduction, and photochemical reduction. These methods can significantly improve the RGO nanostructures conductivity but the highest DC conductivity in RGO is usually less than 1000 S/cm. The low conductivity prevents RGO films from being used as a current collector in LIBs.

Joule heating is a direct processing and modification of carbon nanomaterials. This section introduces that Prof. Yanan Chen and Prof. Liangbing Hu [10] effectively reduced GO in a short time (1 min) at ultra high temperature (about 2750 K) through joule heating, resulting in a high DC conductivity of 3112 S/cm. The DC conductivity is high, which is in connection with the structure of highly crystalline graphene generated by annealing at a high temperature. The high temperature caused by HTS can promote the thermal reduction of GO. Unlike conventional in-furnace heating, joule heating can produce ultra high temperatures at high resistance junctions. Thin (about 4 μm), independent and highly conductive RGO films have a low thin layer resistance of 0.8 Ω/sq and can be used as ultra-light current collectors for LIBs. GO films are thermally reduced in argon (Ar) at 773 K to get pre-anneal GO before joule heating. The pre-annealed RGO has relatively low conductivity and is utilized to touch off joule heating by utilizing an electric current. The RGO film temperature increases with the applied DC bias, monitored by an optical fiber coupled to the homemade device. In the period of joule heating, high temperatures (2750 K) are detected, thermally reducing GO in a few minutes, leaving a highly conductive RGO film. Experimental analysis confirmed that after the high-temperature process, most of the defects were removed (Fig. 3.8a). We use Raman spectroscopy to research the GO structural changes before and after joule heating. Compared with the pre-annealed GO film (Fig. 3.8b), the D-peak of joule-heated RGO film was significantly reduced, and the G-peak of (1600 cm^{-1}) remained at a high intensity peak (Fig. 3.8c), and the I_D/I_G decreased significantly from 1.02 to 0.02.

The low I_D/I_G ratio shows that RGO film treated by joule heating is highly crystalline and its structure changes from amorphous carbon to crystalline carbon. Besides, after joule heating, we can observe a sharp 2D peak (2690 cm^{-1}) with an I_{2D}/I_G ratio of 0.93, which also confirms the highly crystalline structure of RGO films [32, 33]. The results of characteristic peak Raman spectroscopy indicates that highly reduced GO films can be gained by using joule heat generated by current, thus making RGO films with excellent electrical conductivity. The SEM image of RGO film (Fig. 3.8d, e) shows that the structure after joule heating is denser than the original morphology. Before joule heating, the GO film showed that stacked RGO layers of 1.25 μm thickness could be clearly observed. After heating to 2750 K in vacuum

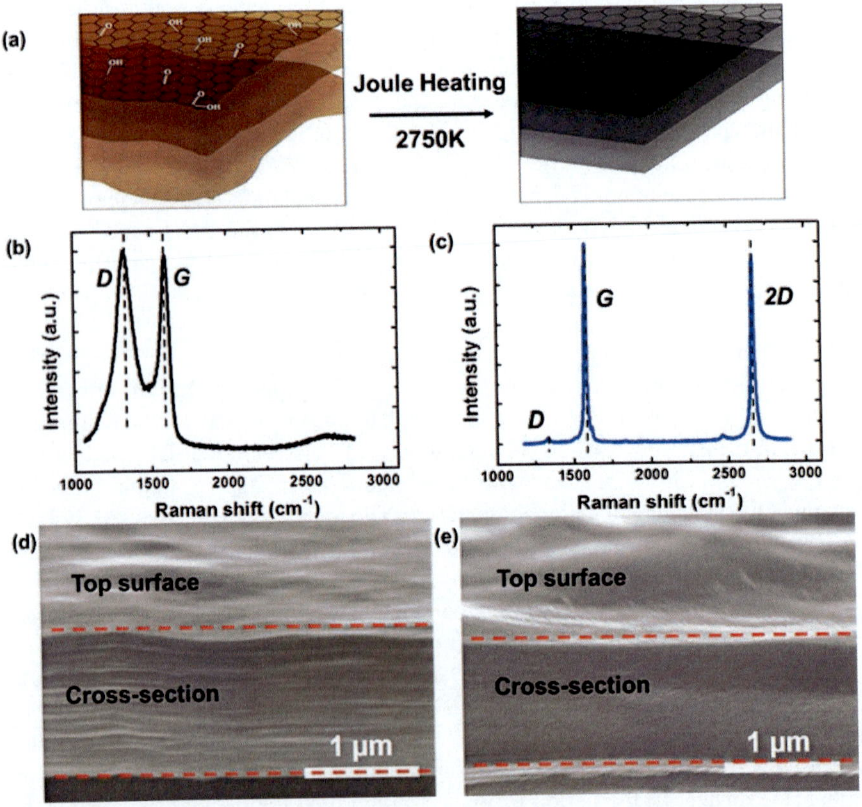

Fig. 3.8 **a** The RGO can be effectively reduced by heating at 2750 K joules for 1 min in vacuum; **b, c** comparison of Raman spectra of RGO films annealed at 773 K and 2750 K for 1 min; **d, e** cross-sectional SEM images of RGO films before and after joule heating

and holding for 1 min, the thickness of RGO film decreases to 1.03 μm, showing a relatively dense section (Fig. 3.8e). This difference in morphology may be due to the high temperature regions "melting" or "welding" together to form highly stacked RGO flakes during joule heating. The high crystallinity (increased conduction within each nanosheet) and dense structure (increased conduction between sheets) of RGO nanosheets simultaneously achieve a high conductivity of 3112 S/cm. The process of joule heating is investigated by utilizing the settings, as exhibited in Fig. 3.9a. The device records the spectrum emitted during joule heating to measure the temperature of the sample. The pre-annealed GO film is hanged above the substrate and both ends of the film are bonded with silver paste to connect the copper electrodes. Perform the joule heating by utilizing direct current (DC) to pre-annealed GO samples (773 K for 1 h) in a vacuum chamber. When an electric current is applied, the RGO film emits light. A spectrometer-coupled fiber with a diameter of 400 μm is placed above the RGO sample. The RGO film and the optical fiber are 15 cm apart to collect light emission. When the voltage is added to 6.55 V/mm, the current density increases immediately from 0.0033 to 0.76 mA/μm^2, about 230 times higher (Fig. 3.9b). The remarkable increase in conductivity may be the high temperature generated by joule heating to remove defects and impurities in the RGO film. In order to characterize the RGO films temperature at a given voltage, the corresponding emission spectra are recorded by spectrometer-coupled optical fibers, as exhibited in Fig. 3.9c. Use the spectral measurement system to record the emission spectra in the wavelength range 340 to 950 nm and then use NIST traceable light sources to calibrate to ensure measurement precision. Then, the emission spectra recorded at high temperature of RGO film are fitted into the blackbody radiation equation to extract the in situ RGO film temperature [34, 35]. Figure 3.9d shows the result of fitting the temperature (T) as a function of the applied electrical power (P). Figure 3.9e shows a series of RGO film images which are taken at different electrical powers. With the input power increasing, the region showing blackbody radiation becomes more pronounced in the image, as shown in the bright region in Fig. 3.9e, indicating that higher temperatures are generated.

Figure 3.10a shows the conductivity increasing by 76 times from 40 S/cm to 3023 S/cm (the sample sizes are 2043 μm long, 212 μm wide, and 1.03 μm thick). In order to get precise values, ten samples were measured and the experimental data were averaged, with a conductivity of 3112 ± 164 S/cm. Figure 3.10b records the relationship between resistivity and reduction temperature during joule heating.

The RGO films are locally heated at high temperatures, particularly in the touched region between RGO nanosheets, which are then "welded" together by defective carbon atoms to form a 3D cross-linked carbon nanostructure. Figure 3.11a, b, respectively shows the RGO film surface morphology before and after joule heating, and each illustration of SEM image shows the enlarged morphology. At 2750 K, the surface of RGO film becomes smoother. The surface topography and cross section showed that the high temperature procedure could remove the irregularity effectively in the structure. XRD and XPS have investigated the microstructure of GO and RGO by thermal reduction at 773 K (pre-annealing) and joule heating (high-temperature annealing at 2750 K). After thermal reduction of GO at 773 K, a (002) characteristic

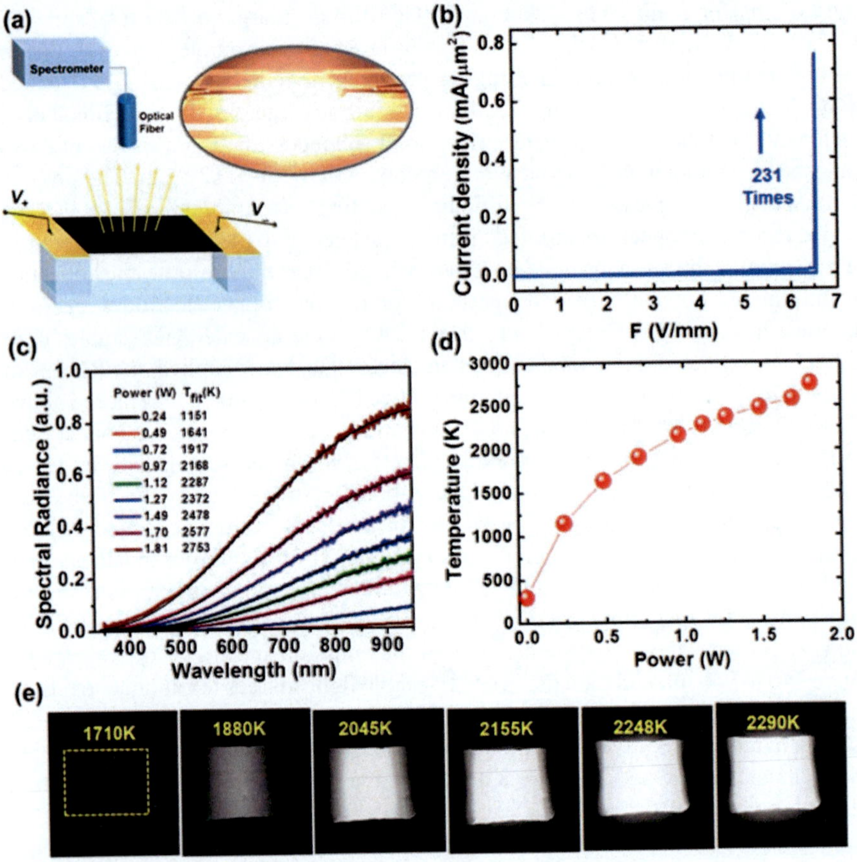

Fig. 3.9 **a** Experimental setup for joule heating experiments in a vacuum chamber with spectrometer and coupled optical fiber; **b** relationship between current density and applied electric field during joule heating of RGO samples pre-annealed at 773 K for 1 h; **c** spectral radiation measurements of the same sample with different electrical power inputs; **d** relationship between installation temperature of RGO film and input electrical power; **e** a set of camera images exhibiting process of the RGO film heating as the electrical power applied increases

peak of graphite appeared at about 26.58° (Fig. 3.11c). After joule heating, the characteristic peaks become narrower and sharper. According to the position of the peak, the RGO interlayer range was obtained, which decreased from 0.369 nm at 773 K to 0.341 nm at 2750 K. The decrease of layer spacing is because the functional groups in RGO films are removed effectively. The narrow peak at 26.58° shows that the uniform stacking space in the RGO of 2750 K is because of the uniform distribution of current in the RGO film. XPS scans were performed on three types of samples, as shown in Fig. 3.11d. The ratio of carbon to oxygen atoms is 2.8 at GO, 3.55 at RGO of 773 K, and 81.64 at RGO of 2750 K, which proves the functional groups are effectively removed. A 1 in. × 1 in. RGO film was demonstrated by solution

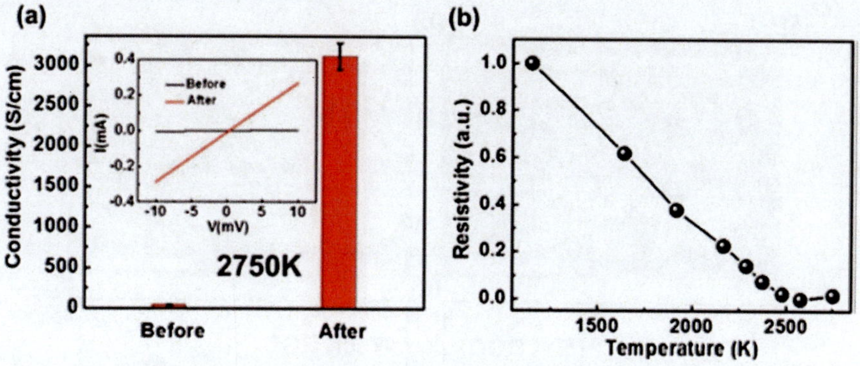

Fig. 3.10 **a** Conductivity of RGO film before and after joule heating to 2750 K; **b** the relationship between the resistance of RGO film and temperature during joule heating

filtration and 1 min high temperature reduction (Fig. 3.11e). RGO films are highly flexible and can be bent to a very little radius without cracking (Fig. 3.11f).

3.3.2 Preparation of Current Collector of Aluminum-Ion Battery by HTS

This section introduces the GO-CNT composite film (RGO-CNT) reduced by joule heating by Yanan Chen/Yida Deng/Wenbin Hu's team [36] and uses it as a supporting flexible current collector in water-based aluminum-ion batteries (AIBs). Because of the surfactant action of GO, GO and CNT can be mixed into GO-CNT evenly and reduced to RGO-CNT rapidly by joule heating at high temperature (2936 K) in a brief time (<1 min). The highly conductive network consists of cross-linked one-dimensional carbon nanotubes sandwiched between densely stacked GO nanosheets. The network could act as a "superhighway" for electronic transmission, turning GO-based materials from insulators into conductors. The conductivity of RGO-CNT is greatly improved to 2750 S cm^{-1} after joule heating, which is 10^6 times higher than that of RGO-CNT. The RGO-CNT obtained can be assembled into flexible AIBs and used as a current collector, showing great electrochemical performance in aspect of excellent rate performance and cyclic stability. In addition, flexible AIBs show extremely high mechanical tolerances (e.g., cutting, piercing folding, or bending).

Figure 3.12 shows the process of joule-induced GO-CNT reduction. During joule heating, CNTs act as electrical and thermal networks. The nanotubes begin to heat in joules. With the time of power supply increasing, the functional groups on GO decrease gradually.

Figure 3.13a shows a photograph of the original GO-CNT film with excellent flexibility and extensibility. A roll-to-roll fabricating system was designed in the laboratory and utilized to joule heating to achieve the reduction of GO-CNT, as

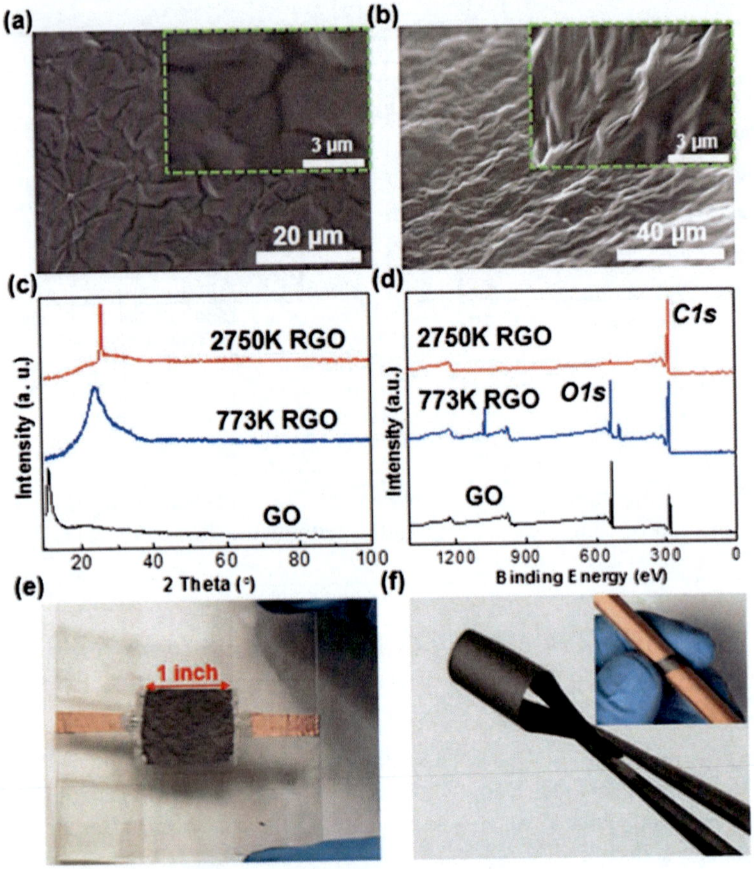

Fig. 3.11 **a** Surface morphology of pre-annealed GO film; **b** surface morphology of RGO films reduced by joule heating; **c** XRD patterns of GO and RGO after annealing at 773 K (pre-annealing) for 1 h, and RGO after annealing at 2750 K by joule heating; **d** XPS scan of the same sample as in **c**; **e** A 1 in. × 1 in. RGO sample was reduced by process of high temperature; **f** High temperature joule-heated RGO has mechanical flexibility

shown in Fig. 3.13b. It can be seen from the continuous snapshot in Fig. 3.13c that the light intensity emitted by the film increases with the increase of the input power, which means an increase in the temperature of reduction.

Figure 3.14a shows a cross section of RGO-CNT, consisting of stacked layers of highly aligned graphene sheets. The thickness of RGO-CNT is 1.4 μm, corresponding to the GO/CNT after joule heating in Fig. 3.13a. The tightly stacked RGO-CNT layer facilitates the transfer of phonons and electrons, thereby improving electrical and thermal conductivity [37]. The magnified cross-sectional SEM image of RGO-CNT (Fig. 3.14b) indicates that CNT is evenly distributed between the RGO layers. Figure 3.14c shows the emission spectrum of a sample which was collected

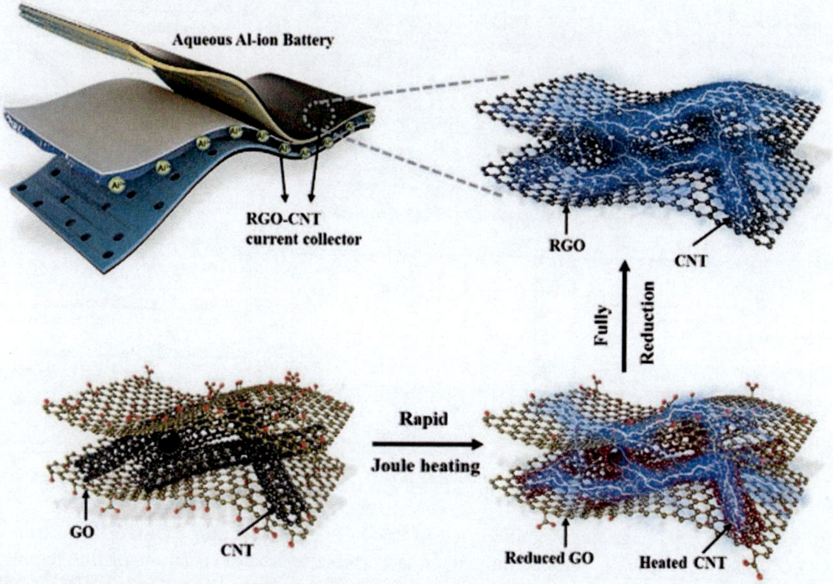

Fig. 3.12 Schematic diagram of GO-CNT reduction induced by joule heating

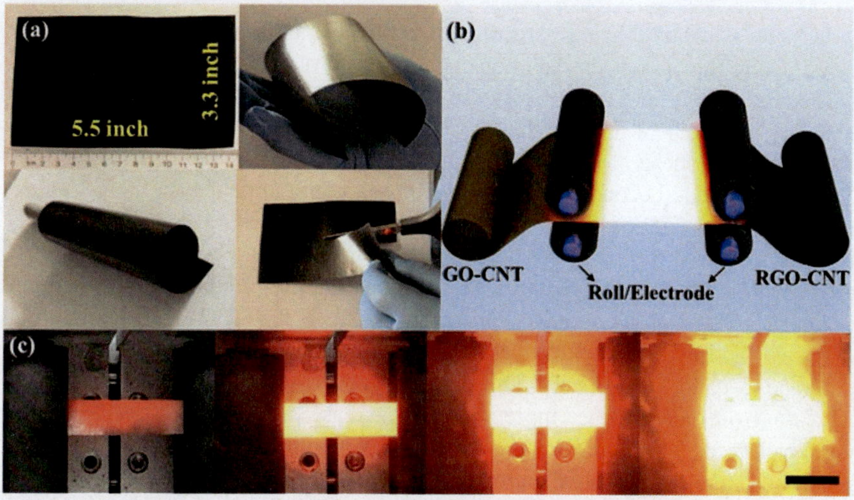

Fig. 3.13 **a** Digital photo of GO-CNT, which reflects the great scalability and flexibility of GO-CNT; **b** schematic diagram of coil-to-coil manufacturing system for continuous reduction of GO-CNT by joule heating; **c** with the increase of input power during joule heating, the light intensity emitted by RGO-CNT film increases

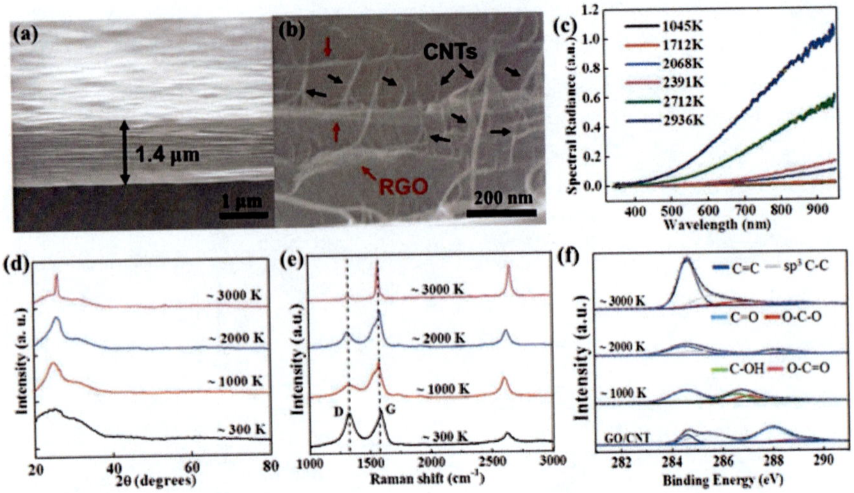

Fig. 3.14 **a**, **b** SEM image and magnified view of RGO-CNT cross section; **c** emission spectrum of RGO-CNT facility, which is suitable for extracting blackbody radiation equation of film temperature; **d–f** XRD, Raman spectra, and high resolution XPS spectra of C 1 s of GO-CNT and RGO-CNT joules anneal at 1000, 2000, and 3000 K, respectively

in the process of joule heating, recorded by a fiber optic spectrometer. The curves were fitted to the blackbody radiation equation to extract the film temperature.

It is noteworthy that the temperature of heat treatment can be as high as about 3000 K. The results of XRD of the samples are exhibited in Fig. 3.14d. The characteristic reflection peak of (002) is at 24.2°. With the increase of heating temperature, the reflection peak becomes sharper. In addition, with the increase of temperature, the characteristic peak shifted from 24.5° to 26.2°, indicating that the interlayer distance decreased from 0.367 to 0.335 nm, and the RGO film was fully graphitized. The reduction of interlayer distance owes to the effective elimination of functional groups in RGO-CNT. The Raman spectra of GO-CNT and RGO-CNT are recorded in Fig. 3.14e, in which the two peaks at 1350 and 1590 cm^{-1} are classified as D-band and G-band, respectively. The intensity ratio (I_D/I_G) of peak D and peak G decreases from 0.98 to 0.14 as the temperature increases, suggesting that joule heating can repair defects in graphene layers effectively [38]. In addition, the G-band and D-band became sharper after joule heating, further demonstrating the recovery of crystal structure [39–41]. Figure 3.14f presents the XPS spectrum of C 1 s. GO-CNT consists of many oxygen-containing functional groups on GO tablets, with a C/O ratio of 0.7. The C/O ratio rises remarkably with the temperature increasing from ≈1000 to ≈3000 K after joule heating. When temperature of the heating reaches ≈3000 K (C/O ratio of 3.3), the peak of $C = C$ becomes apparent, indicating that GO is fully reduced.

Figure 3.15a shows the immediate I-V curve of GO-CNT in the joule heating process. In the early phase of heating, a high voltage (≈80 V) occurs and the current is negligible. Soon, the voltage decreases to ≈8 V. As the current increases further, the voltage increases merely slightly, showing a sharp decrease in resistance. Figure 3.15b shows the change in resistance during the process of joule heating, which decreases from 1.0×10^7 to 11.4 Ω during this period. The temperature-resistance curve of GO-CNT (Fig. 3.15c) shows that the resistance drops from ≈500 Ω to 11.4 Ω with the temperature increasing from ≈1000 to ≈3000 K. Typical I-V curves are shown in Fig. 3.15d. The conductivity of RGO-CNT increased significantly by 2×10^6 times, from 1.11×10^{-3} to 2190 S cm^{-1} after joule heating. The improved conductivity can owe to the reduction of functional groups in GO.

The electrochemical property of AIBs based on RGO-CNT current collector was estimated by cyclic voltammetry (CV) and galvanostatic charge–discharge test. Figure 3.16a exhibits the CV curves of a full aluminum-ion battery based on the positive electrode of CuHCF and the negative electrode of MoO_3 at different scanning rates ranging from 5 to 20 mV S^{-1}. A strong REDOX peak near 0.8/1.0 V can be clearly observed in the figure, which may be attributed to the Al^{3+} embedment/exhalation reaction in the main structure of CuHCF and MoO_3 electrodes. A pair of wide REDOX peaks near 0.4/0.6 V were also noticed. During the oxidation and reduction reactions, the shape of the peak can be maintained well at different

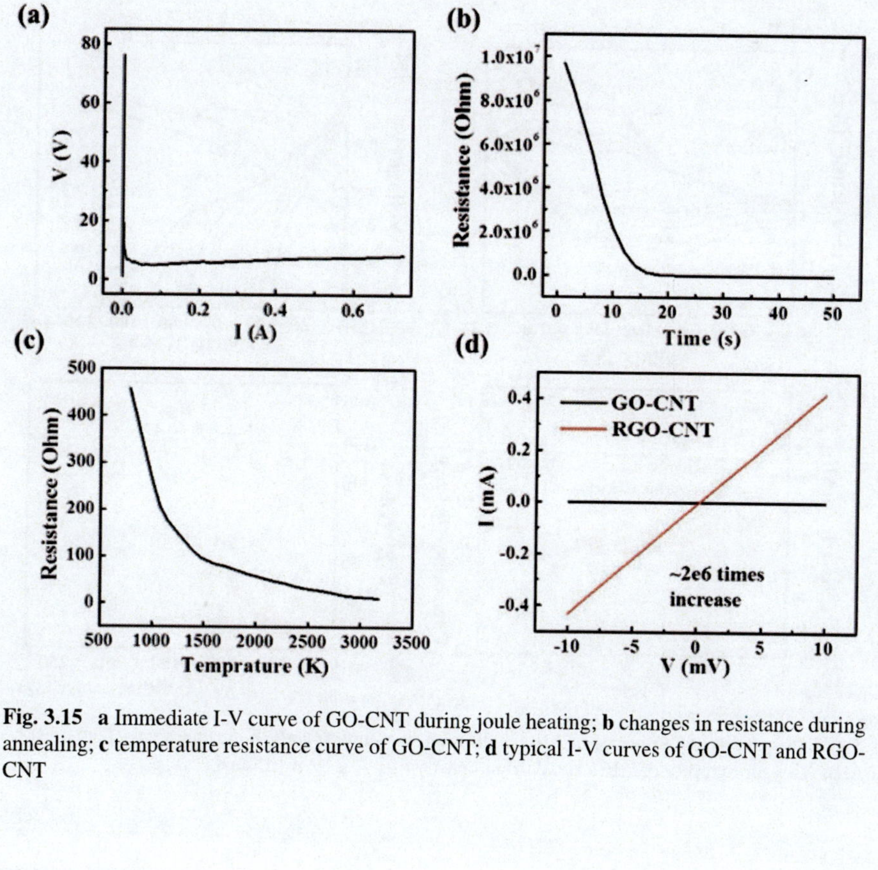

Fig. 3.15 **a** Immediate I-V curve of GO-CNT during joule heating; **b** changes in resistance during annealing; **c** temperature resistance curve of GO-CNT; **d** typical I-V curves of GO-CNT and RGO-CNT

scanning rates, which indicates that the Al^{3+} embedding/deembedding reaction in the main material has excellent reversibility. The charging/discharging test consequences of AIBs are shown in Fig. 3.16b. A long discharge slope was observed with two continuous discharge platforms at about 0.8 and 0.5 V, respectively. They are connected with the Al^{3+} embedment/release process from/into the host material and are very consistent with the REDOX peak of the CV curve. Notably, it provides a discharge capacity of 106.1 mAh g^{-1} at 0.2A g^{-1} current density. In addition, Fig. 3.16c shows the cyclic stability of AIBs at 1 A g^{-1} current density. The initial discharge capacity is 43.1mAh g^{-1},150 cycles and the capacity retention rate is up to 99.5%. In addition, the electrochemical impedance spectroscopy (EIS) test consequences of AIBs are presented in Fig. 3.16d. The small resistance of AIBs suggests the high conductivity of the RGO-CNT collector, which is conducive to the excellent electrochemical performance of water-based AIBs.

To investigate the flexibility of water-based AIBs based on RGO-CNT current collector, the corresponding tests of constant-current charge–discharge were executed under different bending states. As exhibited in Fig. 3.17a, the discharge capacity remains almost constant at bending at 60°, 90°, and 180°, respectively, and a good charging/discharge platform is maintained. This shows that AIBs has excellent performance stability and flexibility. To demonstrate its superior performance as

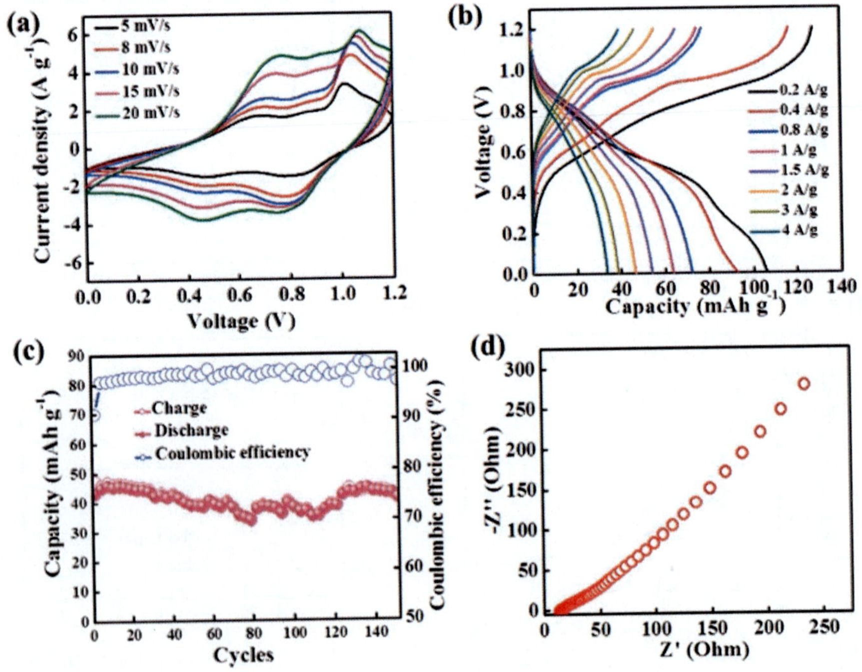

Fig. 3.16 **a** Typical CV curves of AIBs at different scanning speeds; **b** the charge/discharge test of AIBs; **c** cyclic stability of AIBs at a current density of 1 A g^{-1}; **d** EIS test of AIBs

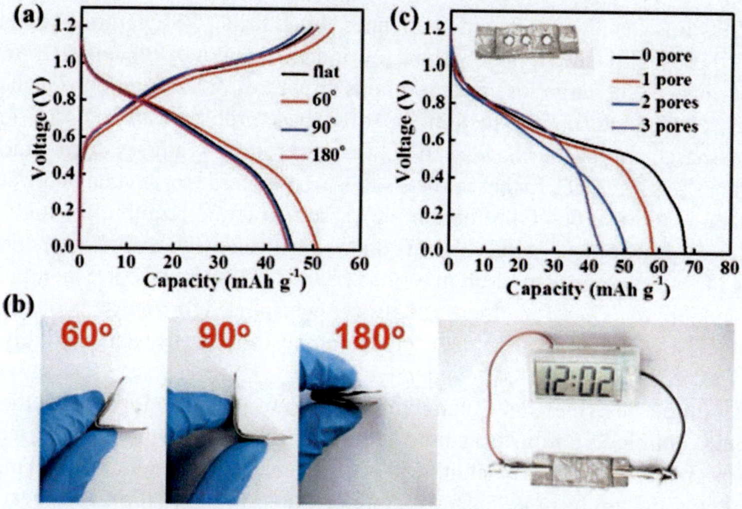

Fig. 3.17 **a** Discharge capacity of waterborne AIBs at 60°, 90°, and 180° bending; **b** an AIBs can continuously light the electronic watch; **c** charge and discharge test of water-based AIBs with several large holes

a power source, an AIBs can continuously light up an electronic meter, as exhibited in Fig. 3.17b. In addition, waterborne AIBs demonstrate high safety to tolerate puncture operations. As shown in Fig. 3.17c, the battery still works normally even with several large holes, with a capacity of 43.8mAh g^{-1}. The water-based AIBs based on RGO-CNT current collector shows high safety, great electrochemical performance, and excellent flexibility, and is a power supply with great promise for wearable energy storage facility in the future.

3.4 Preparation of Solid Electrolyte of Rechargeable Battery by HTS

3.4.1 Preparation of Solid Electrolyte Films by HTS

To avoid the use of flammable liquid organic electrolytes, solid-state electrolytes (SSE) such as garnet based electrolytes have been developed. Solid electrolyte is a key material in all-solid-state batteries. The ideal solid electrolyte should have the advantages of higher room temperature lithium-ion conductivity (higher than 10^{-4} S/cm), negligible electron conductivity (lower than 10^{-8} S/cm), higher electrochemical window (greater than 6 V vs Li^+/Li), stability to lithium metal, environmental friendliness, low cost, easy synthesis, etc. But present approaches of manufacturing such SSE films remain significant challenges. In addition, SSE ceramics manufactured by these approaches typically show low ionic conductivity of about 10^{-8} to 10^{-4} S/cm because of the large loss of amorphous structures or volatile ions (such as Li and Na) deposited during production. To solve these problems, more cost-effective solutions-based approaches have been put forward for the synthesis of ceramic SSE membranes (e.g., garnet). In these processes, SSE is sintered for several hours at high temperatures (600–1100 °C) to obtain the crystal structure required for high ionic conductivity. But because of the volatility of these light elements at high temperatures, prolonged sintering can also result in serious Li and Na losses and corresponding low ionic conductivity. Therefore, there has always been a need for an extensible method for synthesizing ceramic SSE with excellent composition control and crystallinity to realize the necessary high ionic conductivity.

This section describes how Liangbing Hu's team [42] developed a method for producing ceramic SSE films from precursors directly, in which the sintering temperature was greatly increased (reaching 1500 °C), but merely for about 3 s. This method of rapid heating can form a dense structure of polycrystalline film, but because of the short sintering time, the loss of volatile elements is negligible. The method is "Printing and Radiation Heating" (PRH), a printable solution-based technology for the synthesis of ceramic SSE films. In the procedure, the precursor film is printed on a substrate whose thickness is accurately adjusted by controlling the wet thickness and ink concentration. Then the dried precursor film is put in close contact with the radiation heating strip (usually around 1773.15 K) for fast close sintering (Fig. 3.18a). This joule-heated strip passes through the precursor film with a gap of about 1 mm and an overall heating time of a few seconds to finish the sintering procedure, making roll-to-roll manufacturing possible. Figure 3.18b shows a typical PRH sintered $Li_{6.5}La_3Zr_{1.5}Ta_{0.5}O_{12}$ (LLZTO) ceramic SSE film on a single crystal MgO substrate, which has a translucent and dense structure with a surface roughness $< 1\ \mu m$ (Fig. 3.18c).

The precursor powders (Li_2CO_3, La_2O_3, ZrO_2, and Ta_2O_5) were mixed by the method of ball milling and then dispersed in ethanol by ultrasound to produce the precursor (Fig. 3.19a). LLZTO precursor inks have great wettability and fluidity on a variety of substrates such as foil, glass, and ceramics (Fig. 3.19b). Besides, the viscosity and concentration of the precursor ink can be changed to suit different printing technologies, such as spraying and scraping. The precursor inks were deposited by spraying to (i) achieve sintered LLZTO films of various thicknesses (1–100 μm) and (ii) establish patterned films utilizing shadow masks that enable the production of a variety of electrolytes and electrode structures for fabrication of future facility (Fig. 3.19c). For unpatterned and thick films, the precursor slurry can be applied to foil (30 cm × 10 cm) in the surrounding environment using a scraper method (Fig. 3.19d) to produce a uniform and smooth film (Fig. 3.19e). In a representative PRH process, the printed precursor film is rapidly sintered in an atmosphere of argon by moving a joule-heated carbon band over the sample for about 3 s. In order to prove large films on the metal substrates, they manufactured flexible LLZTO/$LiBO_2$ composite films on stainless steel foils with dimensions of 5 cm ×

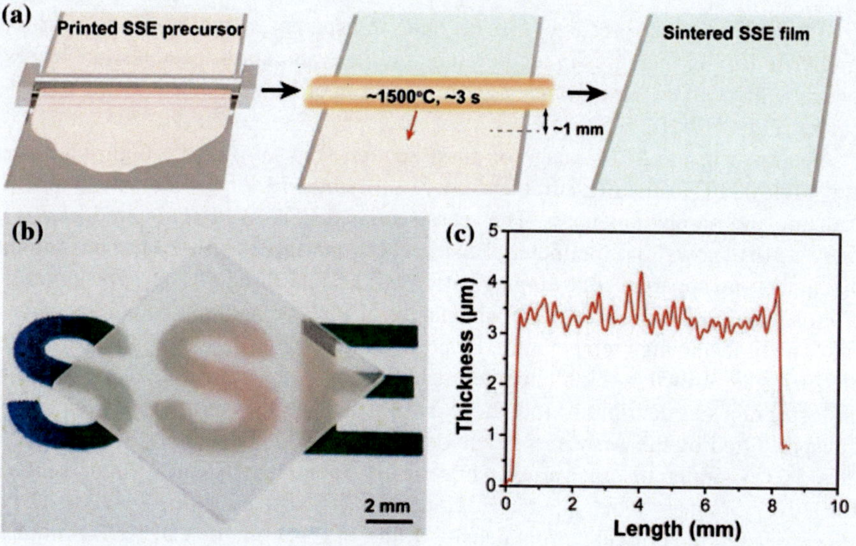

Fig. 3.18 PRH procedure for film synthesis **a** schematic of the film printing technique, which uses a ceramic precursor ink and a rapid sintering process that heats the material to high temperature (1500°C) within ~3 s; **b** the sintered LLZTO garnet film on a single-crystal MgO substrate; **c** the corresponding profilometry curve of the sintered film

2 cm (Fig. 3.20a, b). $LiBO_2$ (30wt.%) was added to reduce the sintering temperature of composite SSE film in order that the stainless steel substrate would not melt. The surface of the LLZTO/$LiBO_2$ film remains conformed and flat during sintering without any apparent cracks or pinholes, indicating great uniformity of the large SSE film produced (Fig. 3.20c–e). The effects of sintering temperature and time on the LLZTO films were investigated by SEM and XRD. At lower sintering temperatures (about 1000 °C), the resulting film has a pure garnet phase, but remains porous even at longer sintering times of 10 s (Fig. 3.19f). At higher sintering temperatures (about 1700 °C), conjunctiva exhibits abnormal grain growth (Fig. 3.19h). The optimal combination of high temperature and short treatment time (1500 °C, 3 s) was found to achieve a dense garnet structure while minimizing lithium loss and side effects (Fig. 3.19g). In order to better understand the evolution of LLZTO precursor films during the process of PRH sintering, the films were sintered at temperatures of about 800–1700 °C and in the time range of 1–180 s. They measured the XRD patterns of the sintered film to characterize the potential phase transitions associated with lithium loss (Fig. 3.19i). In accordance with the XRD pattern, the precursor reacts and forms cubic garnet phase in about 30 s at low temperature of 800–1000 °C. However, the conjunctiva was observed to remain opaque and white at low temperature, indicating the presence of porous structure. Even when the temperature was raised to about 1200 °C and the sintering time was up to 180 s, the porous garnet structure remained unchanged, with a slight loss of lithium observed from the occurrence of $La_2Zr_2O_7$ small peak (Fig. 3.19 I). Nevertheless, when the temperature is increased to 1300 °C

to 1500 °C, the printed precursor film becomes transparent, suggesting a dense structure. With further increases in temperature, the Li evaporation speed also increases fast, as evidenced by the significant phase transition of $La_2Zr_2O_7$ because of serious Li loss (Fig. 3.19i, j).

As shown in the SEM image of cross section (Fig. 3.21a), the high temperature of 1500 °C can offer sufficient energy to sinter LLZTO film with uniform structure and no obvious pores. The cross-sectional EDS of PRH sintered LLZTO (Fig. 3.21b) shows that the material has a clear interface with Al_2O_3, confirming that there is no apparent cross-doping between LLZTO film and Al_2O_3. The absence of cross-doping further proves the advantages of PRH technology combining short sintering time and high temperature. Besides, the process of fast sintering prevented unusual grain growth at high temperatures, as SEM images (Fig. 3.21c, d) showed uniform grain size distribution (about 1.6 ± 0.7 μm). The activation energy of lithium transport fitted by the Arrhenius behavior of lithium-ion conductivity was 0.34 eV (Fig. 3.21e), similar to large garnet, while the ionic conductivity was as high as about 1.0×10^{-3} S/cm at room temperature, comparable to scattered garnet [43, 44]. They also measured the critical current density of the LLZTO film by Li/LLZTO film/Li symmetric cell. The symmetrical cell cycles at a current density of 0.2 to 5 mA/cm^2 for a duration of 10 min (Fig. 3.21f). PRH sintered LLZTO films show a critical current density of 5 mA/cm^2 (Fig. 3.21f), which is one of the highest reported values, even for large garnet SSE [45, 46]. And compared with PRH technology, the traditional sintering process usually takes several hours at > 1000 °C, which results in serious lithium loss. Compared with conventional large particles (e.g., about 1000 μm), thin films (e.g., about 1 μm thick) have a larger specific surface area due to their lower lithium content (Fig. 3.21g).

PRH technology can also be used for layer-by-layer printing and sintering to create layered solid-state batteries. $LiCoO_2$ precursor solution is printed on a thin, fast sintered LLZTO particle and then PRH sintering at about 800 °C (because of the low reaction temperature) for in situ synthesis of $LiCoO_2$ cathode material in about 3 s. $LiCoO_2$/LLZTO/Li solid-state battery is then coated with lithium metal anode on the other side of the particle (Fig. 3.22a). SEM and EDS images of cross section (Fig. 3.22b, c) show that $LiCoO_2$ cathode is uniformly sintered on the surface of LLZTO, with a clear interface, and there is no apparent cross doping with LLZTO garnet (Fig. 3.22c). In order to promote Li transport in the porous $LiCoO_2$ layer and prevent capacity attenuation caused by volume change of the cathode material during circulation, $LiBO_2$ was utilized as a solid binder to mix with the $LiCoO_2$ cathode material. Since $LiBO_2$ can be melted at about 850 °C, we used PRH technology to directly print and sinter the $LiBO_2$ precursor into the porous $LiCoO_2$ layer in only 3 s, thus forming a uniform composite structure (Fig. 3.22d, e). Afterward, the electrochemical properties of all-solid-state $LiBO_2$-$LiCoO_2$/LLZTO/Li cells were characterized.

At 60 °C, the interfacial resistance of PRH sintered cells is as low as about100 Ω·cm^2 (Fig. 3.22f), which is much smaller than that of other sintered all-solid batteries. The charge–discharge curve of PRH sintered battery has a typical $LiCoO_2$ positive platform (Fig. 3.22g), which further proves the successful synthesis of

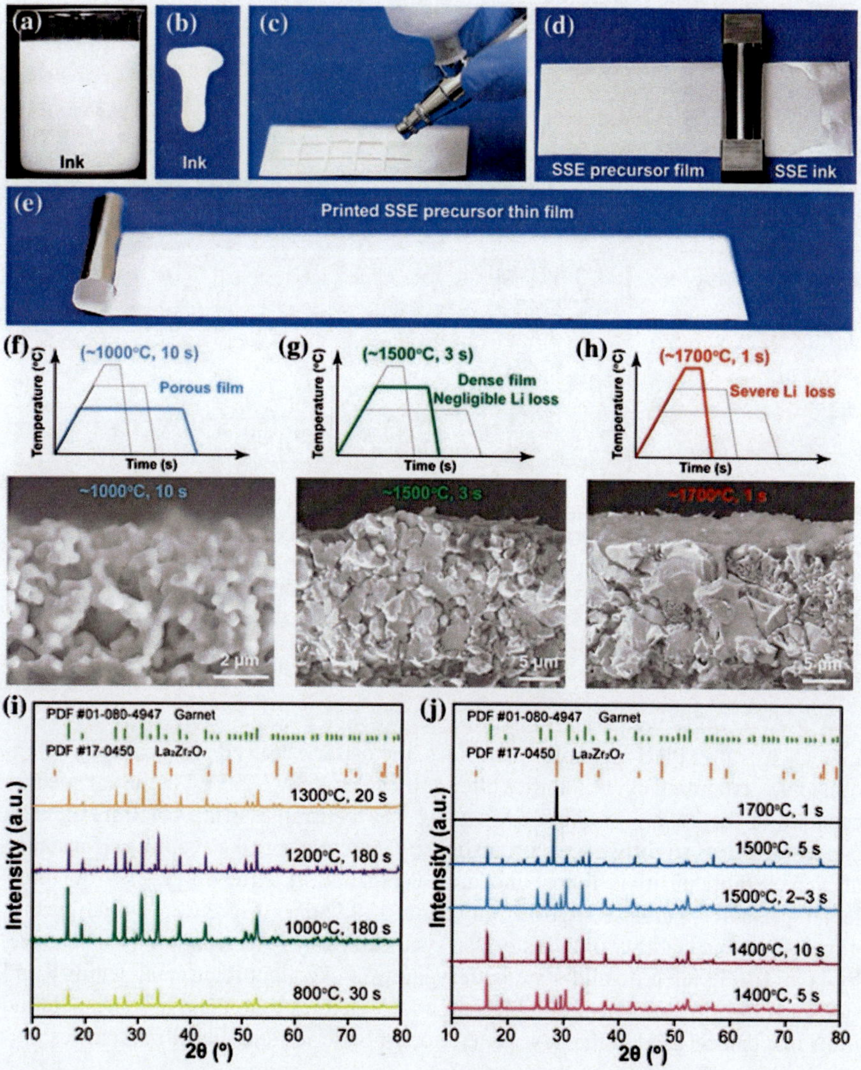

Fig. 3.19 Optimization of printing and sintering conditions for ceramic thin films **a** SSE precursor ink prepared by dispersing the mixed oxide precursors (Li_2CO_3, La_2O_3, ZrO_2, and Ta_2O_5) in ethanol; **b** the SSE ink shows good fluidity; **c** printing the garnet precursor ink by spray coating through a mask; **d** printing the SSE precursor ink by the doctor blade method; **e** the printed SSE film is scalable and flexible; **f–h** schematic and cross-sectional morphology of LLZTO garnet films sintered at different temperatures and times, in which an adequate sintering temperature with appropriate sintering time is necessary to acquire a dense garnet film with limited Li loss and uniform grain size distribution. XRD patterns of LLZTO films sintered at temperatures ranging from **i** 800° to 1300°C and hold times ranging from 1 to 180 s and at **j** 1400° to 1700°C and hold times ranging from 1 to 10 s. a.u., arbitrary units

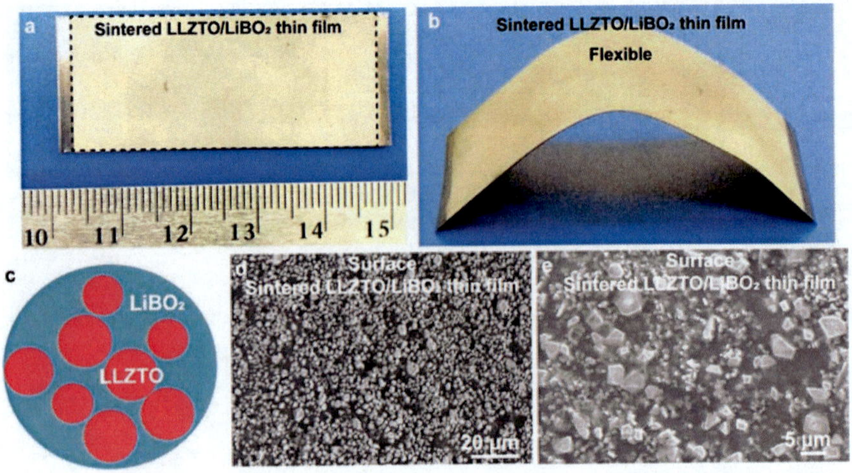

Fig. 3.20 Characterization of PRH sintered flexible LLZO/$LiBO_2$ film **a**, **b** the digital images of the flexible PRH-sintered LLZTO/$LiBO_2$ thin film on a stainless steel foil with size of 5 cm × 2 cm; **c** schematic of the PRH-sintered LLZTO/$LiBO_2$ thin film structure; **d**, **e** surface morphology of the LLZTO/$LiBO_2$ film. 30 wt.% $LiBO_2$ was added to the mixed precursor ink to decrease the sintering temperature of the composite SSE film so that the stainless steel substrate will not melt. The glass $LiBO_2$ uniformly distributed on the grain boundary of LLZTO, forming a dense glass-eramic composite structure without any obvious cracks or pinholes

$LiCoO_2$ by rapid PRH technology. Besides, the battery showed great capacity retention and outstanding cycle stability after 450 cycles (Fig. 3.22h). The initial specific capacity of the battery is about 87 mAh g^{-1} at a current density of 30 mA/g. The capacity reduces slightly as the current density increases, but the cycling performance at each current density is almost unchanged (Fig. 3.22h). After 450 cycles, the interfacial resistance increases slightly, only about 170 $\Omega \cdot cm^2$ (Fig. 3.22f), which further proves the great stability of the interface synthesized by PRH technology. The layer-by-layer PRH sintered solid-state battery showed excellent interfacial stability and stable cycling performance. The PRH approach enables us to mass produce ceramic films that can be used in the next generation of batteries and thin-film facilities.

3.4.2 Clean and Repair Solid Electrolyte by HTS

The cubic garnet phase $Li_7La_3Zr_2O_{12}$ (LLZO) has excellent chemical properties [47] and electrochemical stability to lithium metal, as well as its ability to prevent the growth of dendrite lithium. Therefore, cubic garnet phase $Li_7La_3Zr_2O_{12}$ (LLZO) ceramic solid-state electrolyte (SSE) is one of the most appealing candidate materials for lithium metal battery. However, SSE surface will produce surface contamination and degradation because of many physical or chemical processes at the interface, which is harmful to the ceramic materials performance [48, 49]. The contamination

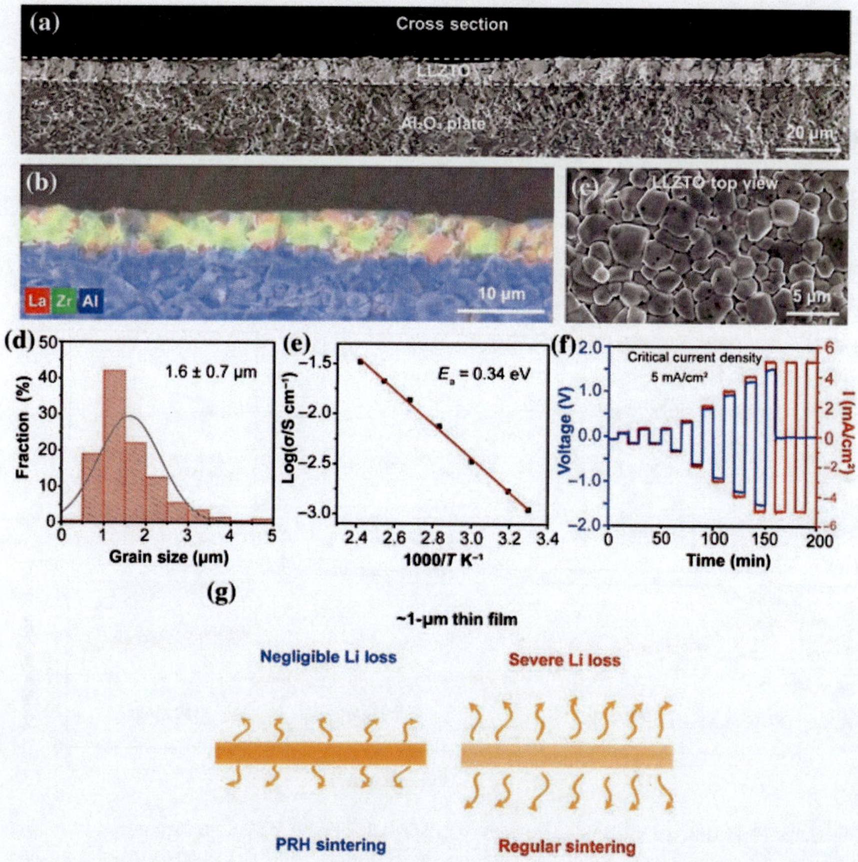

Fig. 3.21 Properties of PRH sintered LLZTO film **a** SEM image of cross section; **b** EDS mapping of LLZTO films sintered on Al_2O_3 substrate; **c** surface morphology of LLZTO films sintered on Al_2O_3 substrate; **d** statistics of grain size distribution of sintered LLZTO films; **e** activation energy of LLZTO films sintered by PRH accords with Arrhenius relation; **f** voltage and current curves of symmetric Li/LLZTO/Li cells with in-plane lithium electrodes for critical current density testing; **g** comparison of lithium loss in films sintered by PRH and conventional approaches

or degradation of the ceramic surface will lead to the deterioration of the interfacial lithium-ion transport property, which will adversely affect the electrochemical performance. Latest researches have displayed that low ionic conductivity, poor electrolyte–electrode interface, and secondary phase contamination at garnet grain boundaries can lead to short circuit of garnet SSE at high current density [50, 51]. Due to the high mobility and reactivity of lithium ions, lithium carbonate (Li_2CO_3) pollution is one of the major causes of these issues. Especially for aging garnet SSE, Li_2CO_3 can accumulate on the surface easily or even at grain boundaries [52], which makes storage of garnet SSE a challenge in actual battery manufacturing.

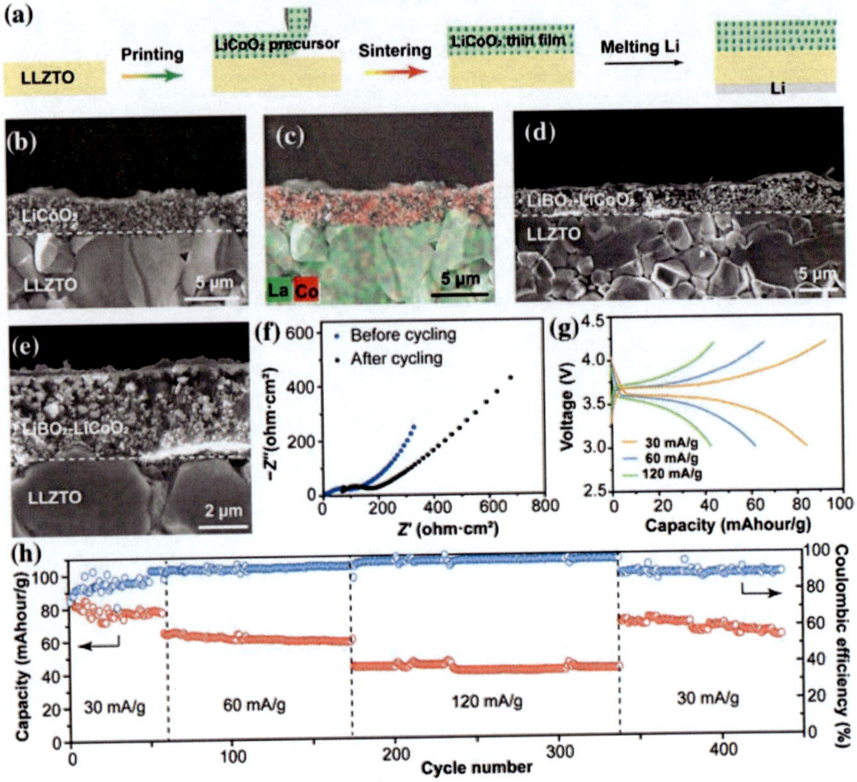

Fig. 3.22 PRH sintered all-solid-state battery $LiBO_2$-$LICoO_2$/LLZTO/Li **a** Printing and sintering processes for manufactured solid-state batteries; **b** SEM images of PRH sintered $LiCoO_2$ positive poles on LLZTO surface and **c** EDS mapping; **d** cross sections of the $LiBO_2$-$LiCoO_2$/LLZTO interface and **e** enlarged SEM images; **f** EIS spectra of all-solid-state batteries ($LiBO_2$-$LiCoO_2$/LLZTO/Li) before and after the 450th cycle; **g** voltage distribution at different current densities for all-solid-state batteries manufactured in situ; **h** cycle performance and Coulomb efficiency of $LiBO_2$-$LiCoO_2$/LLZTO/Li all-solid-state batteries at 60 °C

This section describes how Liangbing Hu's team [53] successfully removed ceramic surface contaminants and repaired aging garnet SSE in less than 2 s through a unique process of thermal pulse. Pulse temperature can reach 1250 °C in 1 s (higher than the decomposition temperature of Li_2CO_3) [54]. The Li_2CO_3 pollutants were completely removed from the surface and grain boundary of SSE of garnet at high temperature. The ultra-short pulse time also avoids lithium evaporation loss and corresponding phase transition successfully. Due to the high processing temperature and inert gas atmosphere, another merit of the thermal pulse procedure is the generation of oxygen vacancies in the SSE of garnet, which acts a great role in the electrochemical performances of ceramics. After heat pulse, the SSE of garnet turned white and the ionic conductivity increased by two times. Benefiting from these effects, thermal pulse garnet SSE exhibits enhanced electrochemical stability.

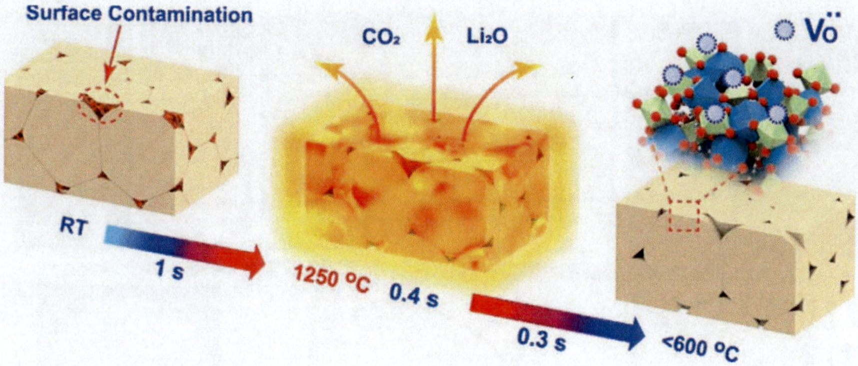

Fig. 3.23 Schematic diagram of SSE fast thermal pulse processing of garnet

The rapid heat pulse process is shown in Fig. 3.23. In this process, gray garnet particles are placed on top of the carbon felt heater. The carbon felt strip is heated by a rapid joule heating, which increases the temperature of the garnet particles to 1250 °C. During the heating process, a slight evaporation of lithium oxide produces oxygen vacancies. As the decomposition temperature of Li_2CO_3 is much higher than that of Li_2CO_3, and carbon mat can also serve on a reducing agent at high temperature, Li_2CO_3 can be completely removed from garnet surface and grain boundary during thermal pulse treatment. In the actual manufacture of garnet based solid-state lithium metal batteries, cleaning effect is important because storage of garnet SSE will inevitably lead to surface generation of Li_2CO_3. The loss of lithium is negligible with a short treatment time, which is crucial for maintaining the phase and ionic conductivity of SSE in garnet. The special fast joule heating approach also makes the roll-to-roll procedure possible for practical application in cleaning garnet SSE.

Figure 3.24a shows the heat pulse processing of SSE of garnet, from which it can be clearly seen that the grain color changes from gray to white. The image and corresponding temperature curve show that the temperature of garnet particles can be heated rapidly from room temperature to 1250 °C in around 1 s, and this high temperature lasts approximately 0.4 s. The natural cooling procedure merely takes about 0.3 s, and the whole procedure takes less than 2 s. The temperature profile was obtained using a VIS–NIR spectrometer utilizing the previous fitting means [55] and can be utilized to detect temperatures above about 600 °C. On account of the heat capacity of garnet particles, the actual temperature of the particles might be slightly lower than the detection temperature of the carbon felt heating strip. Figure 3.24b, c shows that surface contaminants are completely removed after fast heat pulse treatment. SEM images of internal grain boundaries further prove the cleaning effect of fast thermal pulse treatment on SSE of garnet. The untreated grain boundaries of garnet have pollutants and surface (Fig. 3.24d), while the garnet treated by pulse is cleaner and brighter.

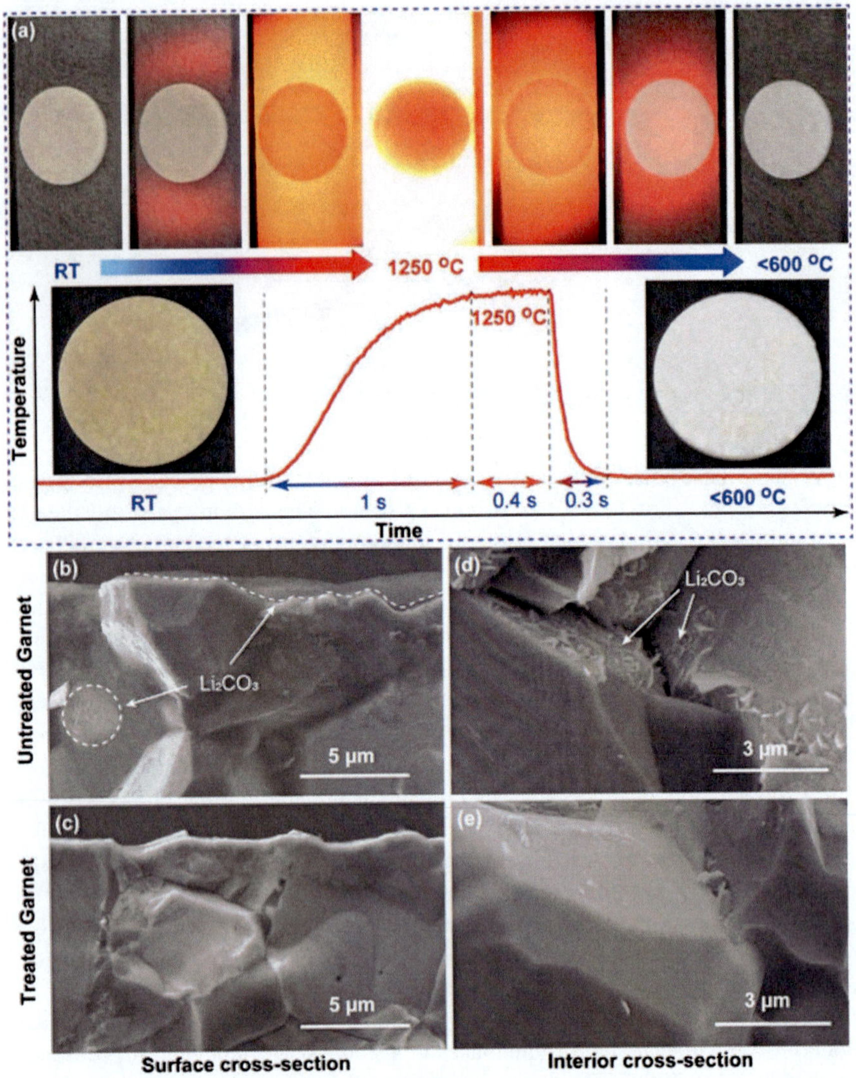

Fig. 3.24 Fast thermal pulse treatment of garnet **a** images of the thermal pulse processing procedure at different temperatures and temperature curves of the procedure; **b** SEM images of cross sections near the surface of garnet particles before and after thermal pulse processing; **d**, **e** SEM image of cross section of internal grain boundary; **d** untreated garnet surface; **e** treated garnet surface

3.5 Application of HTS in Air Battery

Fossil fuels increase CO_2 emissions, leading to global warming and serious climate change. Hence, there is an urgent need to address greenhouse gas emissions. In recent years, there have been more and more studies on the photoelectrochemical reduction of CO_2, especially the development of Li-CO_2 batteries. Rechargeable Li-CO_2 batteries can store energy through the REDOX reaction between CO_2 and carbonate, in which Li_2CO_3 is formed during the discharge reaction and then decomposed during the charging process in the presence of catalysts. However, the discharge product Li_2CO_3 is a wideband gap insulator and thermodynamically stable, so it needs high overpotential to decompose [56]. Therefore, the development of high-performance anode catalysts has become a priority choice for improving Li-CO_2 batteries. But controlling particle size, shape, and dispersion is a huge challenge. At present, the wet chemical route has demonstrated to be simple, inexpensive, and sustainable. However, nanoparticles tend to agglomerate and lose their catalytic activity under high pressure conditions, prolonged reaction time, and the necessary heat treatment [57].

This section describes the synthesis of ultrafine Ruthenium nanoparticles anchored on activated carbon nanofibers (ACNFs) by Liangbing Hu's Team [58] using a transient, one-step HTS method, which can be used as the positive electrode of LI-CO_2 batteries. The original CNF was first synthesized by a simple electrospinning method and subsequently heat-treated under Ar gas at 800 °C for 2 h, leaving a smooth surface without any pores or holes. CNF was then vaporized in a CO2 atmosphere to complete ACNF formation by eliminating unwanted amorphous carbon and effectively generating mesoporous structures. There are more defect sites in ACNF after CO_2 gasification than in CNF, as the ID/IG ratio increases from 0.97 (CNF) to 1.03 (ACNF) [32]. In addition, the increased surface area and pore volume of ACNFs compared with CNFs indicated that ACNFs could provide more adsorption sites for ruthenium salts and deposition sites for ultrafine Ru nanoparticles. That is, $RuCl_3$ can merely absorb on the surface of CNF and generate relatively large Ru nanoparticles (9.0 ± 2.5 nm) on its exterior (Fig. 3.25a). It should be noted that the unique structure of ACNF should result in high dispersion of Ru nanoparticles, with small size and narrow size distribution (4.1 ± 0.9 nm), as exhibited in Fig. 3.25b. Through TGA, the content of Ru in Ru/CNF and Ru/ACNF was 10.7 wt.% and 18.1 wt.%, respectively, indicating a high Ru load in ACNF.

BET surface area of CNFs and ACNFs was 63.18 and 379.91 $m^2\ g^{-1}$, respectively, while t-plot micropore volume was 3.85×10^{-3} and $136.74 \times 10^{-3}\ cm^3\ g^{-1}$, respectively. These results indicate that the increased surface area and pore volume of ACNFs after CO_2 activation can offer more adsorption sites for ruthenium salts and deposition sites for ultrafine Ru nanoparticles, not only on the surface of ACNFs, but also within ACNFs. The porous structure of ACNFs promotes carbon dioxide transport, electrolyte penetration, and lithium-ion transport, ultimately improving the electrochemical property of Li-CO_2 cells by increasing reversibility and circulation.

Rapid synthesis of Ru/CNF and Ru/ACNF electrodes is achieved by ultra-fast method of HTS, in which the corresponding temperature is controlled by the applied current. First, the $RuCl_3$ precursor was loaded onto CNF and ACNF networks, and then decomposed into Ru metal by applying an 800 mA current of 55 ms, which then formed nanoparticles during rapid cooling. Figure 3.26a shows photographs of ACNF-$RuCl_3$ films before and during treatment of HTS. The full spectrum in the

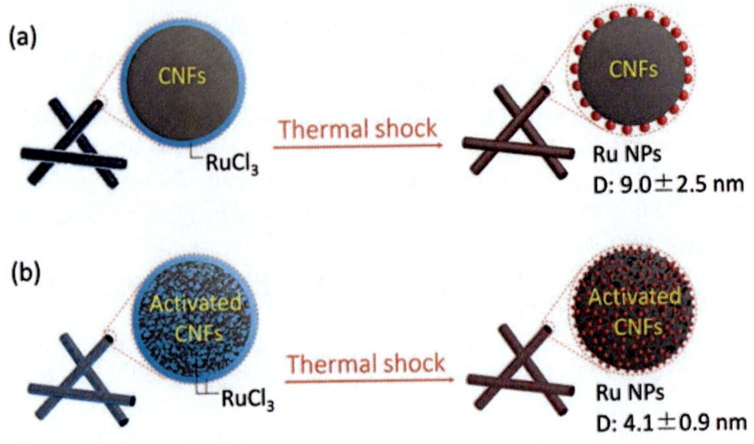

Fig. 3.25 Schematic diagram of Ru nanoparticles formed on **a** primitive CNFs, **b** activated CNFs (ACNFs) by instantaneous HTS

wavelength range of 464–867 nm (Fig. 3.26b) was collected, and temperature and time curves during HTS were extracted by fitting Planck's law [59, 60]. Figure 3.26c shows the temperature and time curve. The temperature rises rapidly to 1600 K after 44 ms, and the temperature cools significantly by removing the current within 40 ms. The decomposition temperature of 1600 K is higher than that of $RuCl_3$, which enables the material to form Ru nanoparticles.

To study the microstructure and morphology of Ru/CNF and Ru/ACNF electrodes, SEM and TEM were carried out. SEM and TEM images of Ru/CNFs in Fig. 3.26d, e show that Ru nanoparticles are distributed on CNFs with a particle size distribution ranging from 3 to 16 nm and an average size of 9.0 ± 2.5 nm (Fig. 3.26f). Ru/ACNF materials were studied using SEM and TEM images (Fig. 3.26 g–j). The average size of uniformly anchored Ru nanoparticles on ACNF matrix is 4.1 ± 0.9 nm and shows a narrower and more uniform distribution of particle size (from 2 to 6 nm) than Ru/CNF, as exhibited in Fig. 3.26i. HR-TEM images of Ru/ACNFs show crystal plane spacing of 0.205 nm and 0.214 nm, which corresponds to the (101) and (002) crystal planes of hexagonal Ru, showing great crystallinity of Ru nanoparticles (Fig. 3.26k). The SAED pattern in Fig. 3.26i exhibits two Bragg reflections of hexagonal Ru (101) and (002), which are in line with HR-TEM and XRD results. Figure 3.26m and n shows SEM images of typical cross sections of Ru/CNF and Ru/ACNF films, suggesting differences between materials. Although the cross section of Ru/CNFs is smooth, porous, and rough structures can be observed inside Ru/ACNFs. Due to the porous structure of ACNFs, Ru nanoparticles can be deposited not only on the fiber surface, but also in the interior of individual fibers in the Ru/ACNF electrode successfully. The corresponding EDX spectra and elemental composition (Fig. 3.26o) showed that only Ru nanoparticles were observed on the surface of Ru/CNFs, further suggesting the important role of porous materials. Based on these results, it is inferred that the utilization of ACNF as a positive substrate offers plenty of sites and defects for the

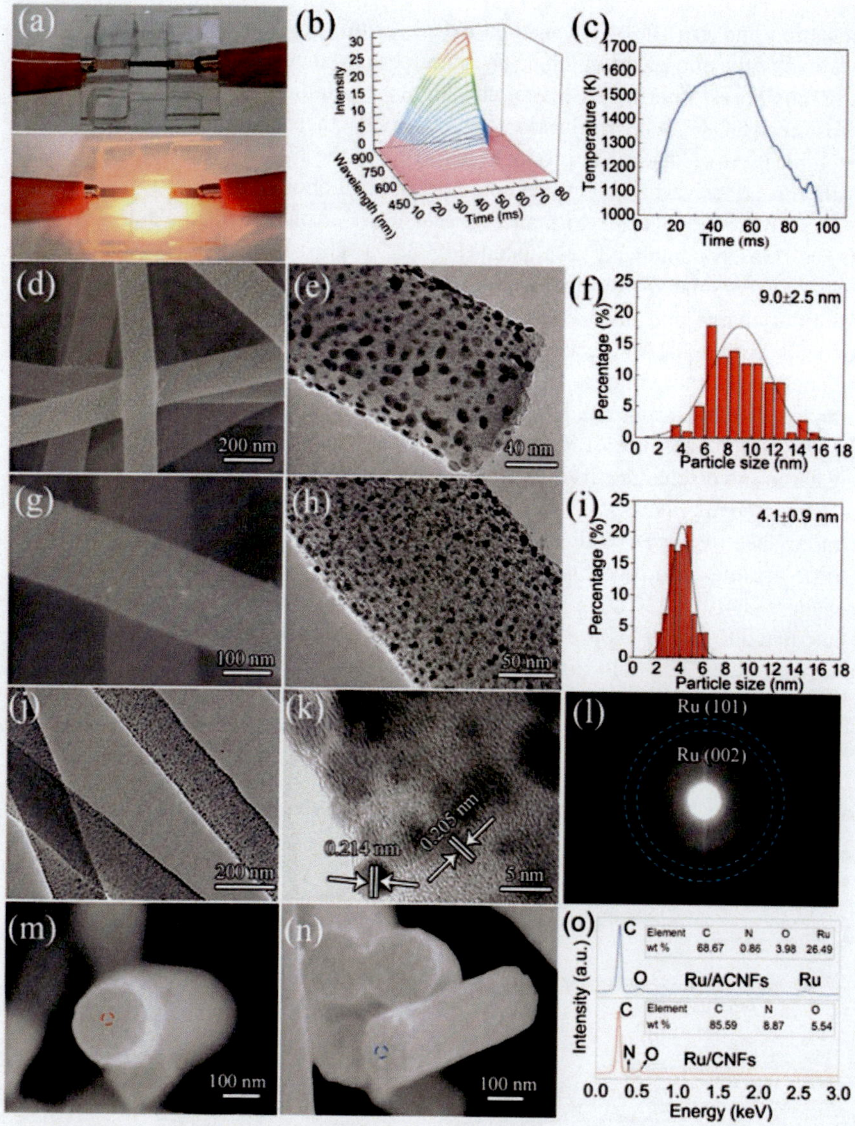

Fig. 3.26 HTS and characterization of Ru nanoparticles immobilized on carbon nanofibers **a** optical photographs of independent ACNF-$RuCl_3$ films on glass supports prior to treatment of HTS; **b** emission spectra in the wavelength range of 464–867 nm conform to Planck's law of gray body radiation; **c** extracted temperature and time curves of 55 ms treatment of HTS; **d–f** SEM, TEM, and Ru particle size distribution of Ru/CNFs show that the particle size distribution is 9.0 ± 2.5 nm; **g** SEM images; **h**, **j** TEM images; **k** HR-TEM images; **i** the Ru granularity distribution of Ru/ACNFs, suggesting the ratio of Ru/CNF; **l** SAED mode of Ru/ACNFs, suggesting the polycrystalline structure of Ruthenium nanoparticles; **m** SEM images of cross sections of Ru/CNF electrode; **n** SEM images of cross sections of Ru/ACNF electrode; **o** EDX spectra and elemental composition of marked positions in SEM images of cross section (insert chart), demonstrating that Ru nanoparticles can be deposited in the internal structure of ACNF successfully

nucleation and growth of Ru nanoparticles, resulting in smaller particle size and coverage of the entire carbon fiber. These properties, such as ultrafine nanoparticles, numerous pores, interlinked fibers, and highly crystalline structures, facilitate the diffusion of CO_2 and the infiltration of electrolytes, thereby improving the catalytic decomposition of discharge products involved in the electrochemical cycle. As a result, this adhesive-free electrode could be an ideal choice for Li-CO_2 batteries.

The electrochemical performance of Ru/ACNF in the positive electrode of Li-CO_2 cell was systematically evaluated (Fig. 3.27). The specific capacity and current density are calculated from the overall mass of the electrode. Figure 3.27a exhibits the initial charge and discharge curves of Ru/ACNF and Ru/CNF electrodes at a current density of 0.1A g^{-1}, with a specific capacity of 1000 mAh g^{-1}. The discharge and charge terminal voltages of Ru/ACNFs are 2.80 V and 4.15 V, respectively. In comparison, the overpotential of Ru/CNFs is higher because of the large size of Ru particles and the absence of Ru particle scales in CNFs.

Charge and discharge curves of Ru/ACNF electrode at different winding numbers are presented in Fig. 3.27b. According to the previous XRD, Fourier transform infrared spectroscopy (FTIR) and differential electrochemical mass spectrometry (DEMS) studies and investigations on discharge products, the proposed discharge procedure is $4Li^+ + 3CO_2 + 4e^- \rightarrow 2Li_2CO_3 + C$ [61, 62]. The theoretical voltage is calculated to be 2.8 V based on this reaction. The results exhibits that the Li-CO_2 battery operates at a high discharge voltage over 2.76 V (vs. Li^+/Li), which is in line with the theoretical value. Even after 50 cycles, the corresponding overpotential between the charging and discharging terminal voltages can remain at about 1.43 V,

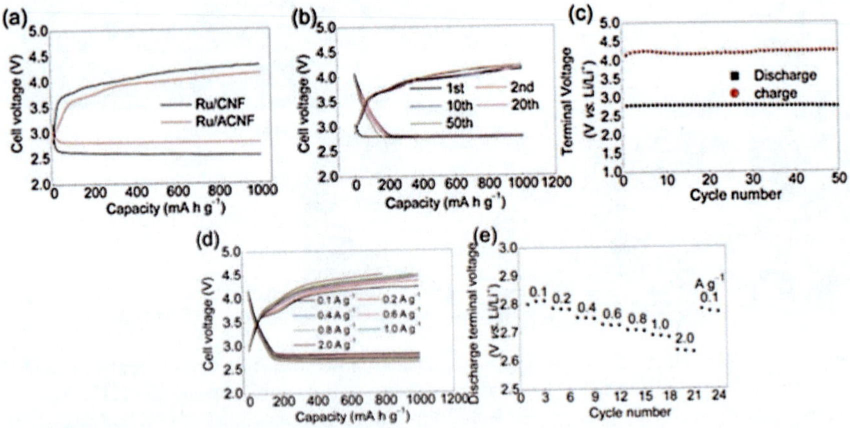

Fig. 3.27 Electrochemical performance of Ru/ACNF anode in Li-CO_2 cell **a** charge–discharge curves of Ru/CNFs and Ru/ACNFs at 0.1 A g^{-1} current density, with a specific capacity of 1000 mA h g^{-1}; **b** charge and discharge curves of Ru/ACNF electrode at the 1st, 2nd, 10th, 20th, and 50th cycles when the current density is 0.1A g^{-1}; **c** the corresponding discharge and charging terminal voltage during the cycle; **d** charge and discharge curves of Ru/ACNF electrode at different current densities; **e** discharge termination voltages of Ru/ACNF electrodes at different current densities

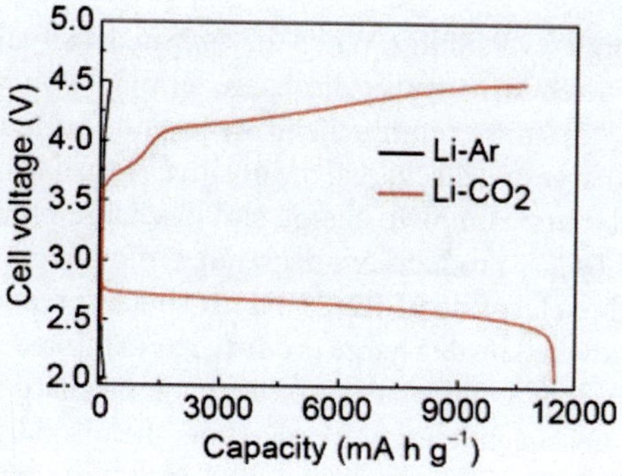

Fig. 3.28 Charge and discharge curves of Ru/ACNF anode at 200 mAg^{-1} current density in Li-Ar and Li-CO_2 cells

indicating excellent cycle stability, as exhibited in Fig. 3.27c. The discharge terminal voltage of Ru/ACNF is higher than that of the positive electrode of ACNF, and the charging terminal voltage is lower than that of ACNFs, suggesting that ultrafine Ru nanoparticles can not only catalyze the generation of Li_2CO_3, but also the precipitation of CO_2 as a catalyst. They further examined the discharge and charging curves of the positive Ru/ACNF electrodes in Li-Ar and Li-CO_2 cells, as exhibited in Fig. 3.28.

The capacity of the positive Ru/ACNF electrode obtained in argon is negligible, indicating its ultra-low capacitance contribution. However, the positive electrode of Ru/ACNF shows a long discharge platform at 2.72 V, and the corresponding initial discharge capacity is reaching 11,495 mAh g^{-1} in the presence of CO_2. When the battery is charged back to 4.5 V, 10,715 mAh g^{-1} reversible charging capacity can be obtained. These results indicate that ultrafine Ru nanoparticles can not only catalyze the generation of Li_2CO_3, but also the formation of CO_2. Based on the electrochemical performance in Fig. 3.27a–c, we can attribute the high performance of Ru/ACNF positive electrode in Li-CO_2 cells to porous ACNF and superfine Ruthenium catalysts.

They thought that this phenomenon can be accounted for the fact that the porous and defect-rich structure of ACNF can offer more adsorption sites for ruthenium trichloride ($RuCl_3$) salt and thus for deposition of ultrafine Ru nanoparticles. In addition, the porous structure of ACNF substrates promotes the infiltration of electrolyte, the transport of CO_2, and the diffusion of Li^+. The holes also provide enough space to stockpile discharge products. At last, due to the defect of ACNF substrate, the growth of ultrafine Ru nanoparticles promoted excellent catalytic activity and reversible reaction between Li_2CO_3 and carbon. Figure 3.27d shows the charge/discharge curves of Ru/ACNF electrode in Li-CO_2 battery at different current densities. As the current density increases from 0.1 to 2.0 A g^{-1}, a small increase in overpotential implies excellent rate performance. When the current density is 0.1, 0.2, 0.4, 0.6, 0.8, and 1.0 A g^{-1}, the discharge terminal voltage of Ru/ACNF electrode in Li-CO_2 battery is 2.80, 2.78, 2.75, 2.72, 2.70, and 2.68 V, respectively. Even at a high current density of 2.0 A g^{-1}, the discharge terminal voltage remains stable at 2.64 V, indicating that the rate performance is good because of the porous structure of the positive Ru/ACNF electrode. In the meantime, Li-CO_2 cells with positive Ru/ACNF electrodes show high reversibility. When the current density reduced to 0.1A g^{-1} after cycling at high current density, the discharge terminal voltage recovered to 2.77 V (Fig. 3.27e).

To further comprehend the reaction principle of Li-CO_2 battery after the process of charge and discharge, the positive electrode of Ru/ACNF was analyzed by SEM after the first complete charge and discharge process. After the first complete discharge, Li_2CO_3 products with granular form were deposited on the surface of the positive electrode of Ru/ACNF. It should be noted that the electrode is not completely covered by discharge products, as exhibited in Fig. 3.29a. TEM images in Fig. 3.29b further confirm the presence of discharge products with granular form. In the subsequent charging procedure, the discharge products nearly disappeared, and the Ru/ACNF electrode returned to its original form, indicating that the battery had good reversibility and rechargeability (Fig. 3.29d). As TEM images demonstrate, Ru nanoparticles are also firmly fixed to ACNF after cycling (Fig. 3.29e), laying the foundation for long cycle life battery systems. Corresponding schematic diagrams of discharge and charge state are exhibited in Fig. 3.29c, f, respectively.

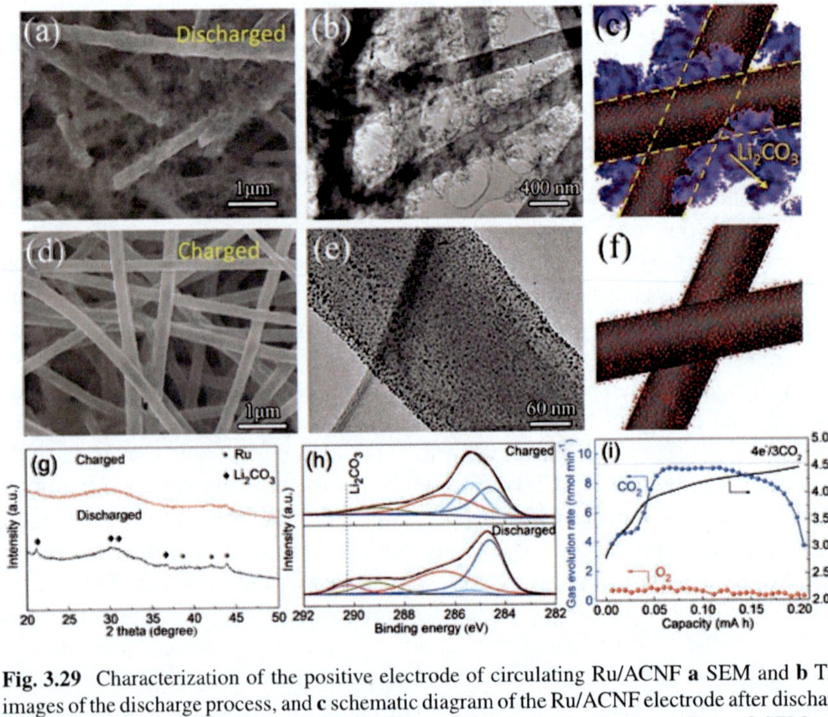

Fig. 3.29 Characterization of the positive electrode of circulating Ru/ACNF **a** SEM and **b** TEM images of the discharge process, and **c** schematic diagram of the Ru/ACNF electrode after discharge, showing that the discharge products of Li_2CO_3 are deposited on carbon nanofibers; **d** SEM and **e** TEM images of the charging process, and **f** the schematic diagram of Ru/ACNF electrode after the first charging, showing that the discharge products of Li_2CO_3 could be decomposed, indicating that there was a reversible reaction between CO_2 and Li_2CO_3; **g** XRD patterns of Ru/ACNF electrode after full discharge to 2.0 V and full charge to 4.5 V; **h** XPS spectrum; **i** gas escape speed of the positive electrode of Ru/ACNF during charging at a current of 20 mA and a capacity of 0.2 mAh

In order to further study the reversibility and evolution of the discharge products and electrochemical reactions, ex situ XRD, FTIR, and XPS measurements were performed on the positive electrode of Ru/ACNF during the full discharge and charging stages. After charging, all diffraction peaks related to Li_2CO_3 disappeared with CO_2 precipitation reaction, indicating that Li_2CO_3 was fully decomposed and Ru nanoparticles reappeared on the surface of ACNF. As expected, characteristic peaks of Ru nanoparticles can still be observed clearly, suggesting that there are still many active sites that facilitate the reversible transformation between Li_2CO_3 and CO_2. This result is also verified by high resolution XPS of C 1 s spectrum. During the subsequent charging process, the 290.2 eV peak value of the C = O bond in Li_2CO_3 was significantly reduced (Fig. 3.29h). Besides, in situ DEMS measurements were used to monitor gas escape from the positive electrode of Ru/ACNF during battery charging, as exhibited in Fig. 3.29i. Only CO_2 gas was detected, and the charge/mass ratio was determined to be $4.1e^-/3CO_2$, close to the theoretical value based on the reversible reaction $4.1e^-/3CO_2$: $4Li^+ + 3CO_2 + 4e^- \rightarrow 2Li_2CO_3 + C$. The DEMS results are consistent with earlier researches, although there is no confirmation that carbon emissions are a complete source of CO_2 generation. These results indicate that ultrafine Ru nanoparticles can promote the generation of Li_2CO_3 and its reversible decomposition into CO_2 during discharge and charging.

This transient in situ HTS synthesis approach combined with CO_2 gasification gives an easy and effective way to produce highly dispersed superfine metal nanoparticles on carbon substrates. This study provides a method with great promise for the development of Li-CO_2 batteries with outstanding cycling and multiplier performance, which can also be used for other catalytic and applications of renewable energy storage. In this section, Liangbing Hu's team [63] proposed an effective strategy which combines 3D printing with HTS processing for the design of thick electrodes for the synthesis of high-energy density Li-CO_2 batteries. The ultra-thick (up to ≈0.4 mm) positive electrode was manufactured in two steps: Firstly, the GO frame was constructed by 3D printing, and then reduced in argon (Ar) for 1 h to obtain RGO skeleton; Secondly, after immerging the frame in a $NiCl_2$ solution, ultrafine Ni nanoparticles were fixed to RGO by rapid (54 ms) and HTS (1900 K) heating. Because of the design of thick electrode and uniform distribution of ultrafine catalyst nanoparticles, the ultrafine positive electrode presents a low overpotential of 1.05 V and a high area capacity of 14.6 mA h cm^{-2} at 100 mA g^{-1}.

Figure 3.30a shows the temperature and time curves of RGO frame and $NiCl_2$ during HTS treatment. After 4 ms, the temperature increased to ≈1900 K gradually, which was sufficient to decompose $NiCl_2$ into Ni nanoparticles and then cooled rapidly within 20 ms by heat conduction and radiation. Field emission scanning electron microscopy (FESEM) showed that the RGO frame remained unchanged even after HTS treatment (Fig. 3.30b). They further analyzed the distribution and fine structure of Ni nanoparticles in the RGO framework. Figure 3.30c shows stacked RGO filaments (the part marked in yellow), approximately 500 μm in diameter, suggesting that the frame can maintain its original configuration after HTS. As shown in Fig. 3.30d, the corresponding high magnification image shows that a large number of Ni nanoparticles are uniformly distributed on the top of the entire frame.

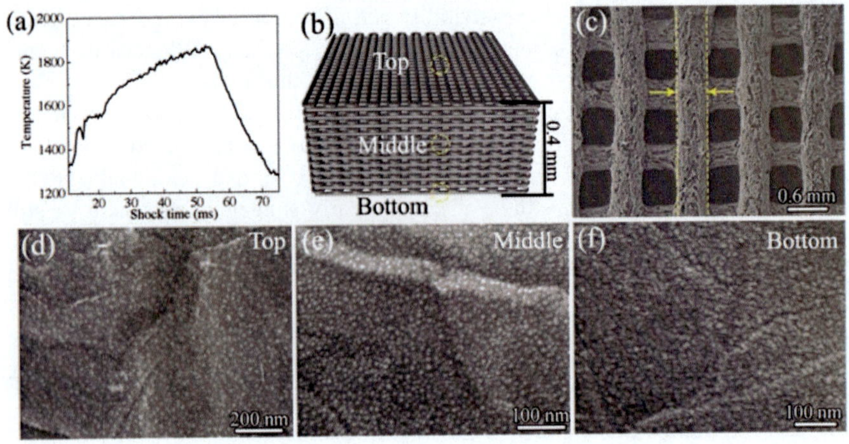

Fig. 3.30 **a** Temperature and time curves of the RGO frame containing $NiCl_2$ during HTS treatment, e.g., spectral measurements; **b** schematic diagram of Ni/RGO framework; **c** a low-power FESEM image at the top of the electrode shows the interconnect mesh and porous structure of the material; **d–f** high magnification FESEM images of Ni/RGO skeletons at top, middle and bottom after HTS treatment, respectively

Ni nanoparticles were monodispersed and fixed on RGO sheets with an average size of about 5 nm. At the same time, Ni nanoparticles can also be observed in the middle and bottom of the RGO frame, as shown in Fig. 3.30e, f. These consequences exhibit that Ni nanoparticles can be anchored on RGO sheets uniformly, which is owed to the combination of the porous structure of the frame and the rapid kinetics of HTS. This highly porous network with many cavities facilitates impregnation of $NiCl_2$ solution in RGO frames. Meanwhile, the high thermal conductivity and ultra-fast HTS method of RGO can uniformly decompose and grow Ni nanoparticles.

Next, they tested the electrochemical properties of Ni/RGO as a positive electrode in Li-CO_2 cells. Figure 3.31a exhibits the charge and discharge performance at 100 mA g^{-1} current density with a specific capacity of 1000 mA h g^{-1}. The electrode showed stable cycling performance and apparent potential platform after 100 cycles (Fig. 3.31b). In addition, the overpotentials calculated by the charge–discharge end voltmeter were 1.05, 1.25, 1.17, 1.18, 1.17, 1.21, 1.23, and 1.25 V for the 1st, 10th, 20th, 30th, 50th, 60th, 80th, and 100th laps, respectively. These results indicate that Ni/RGO framework has excellent catalytic stability in Li-CO_2 cells.

Figure 3.31c exhibits the discharge and charge curves at different current densities. All discharge and charging voltage platforms can be seen even if the current density increases from 200 to 1000 mA g^{-1}. At different current densities, the discharging end voltages were 2.81, 2.73, 2.67, 2.63, 2.60, 2.57, 2.48 V, and the corresponding over-potentials were 1.36, 1.56, 1.67, 1.75, 1.81, 1.86, 2.01 V, respectively (Fig. 3.31d). The discharge voltage is significantly improved, and the rate performance is better than earlier studies on nickel based positive electrodes in Li-CO_2 cells, because the

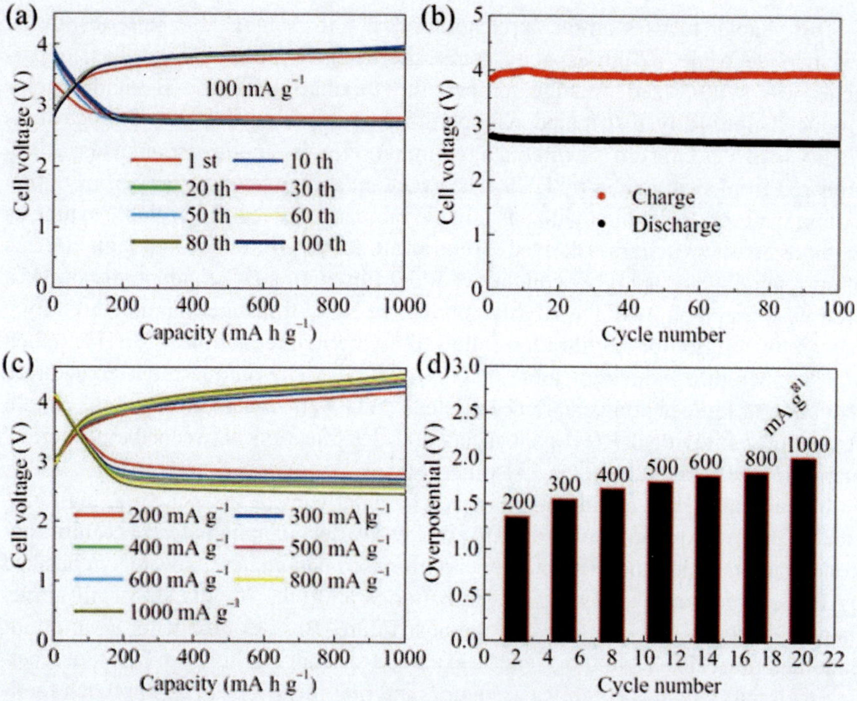

Fig. 3.31 **a** Discharge and charge curves of Ni/RGO skeleton as positive electrode at 100 mA g^{-1} current density; **b** discharging and charging end voltage during the cycle; **c** discharge and charge curves; **d** the corresponding overpotential of the positive electrode at different current densities is limited to 1000 mAh g^{-1}

maximum rate reported in the study is 200 mA g^{-1} and the corresponding overpotential is 1.9 V. This is the first time that 3D printing technology has been combined with HTS method to produce an ultra-thick (up to ≈0.4 mm) positive electrode composed of a Ni/RGO skeleton for Li-CO_2 batteries. As the positive electrode of the Li-CO_2 cell, the Ni/RGO skeleton provides a high area capacity of 14.6 mAh cm^{-2} during initial discharge. In addition, the cell exhibits stable cycle performance, high-rate performance up to 1000 mA g^{-1}, and low charge/discharge overpotential of 1.05 V. This work provides a convenient and effective strategy for designing thick electrodes.

3.6 Summary

Based on HTS technology, the reaction temperature can reach ultra-high levels (>3000 K) instantly in a very short time, and the heating/cooling speed is 10^5 K s^{-1}. In recent years, this technology has been widely used in the preparation of many kinds of functional nanomaterials, particularly in energy conversion and storage.

This chapter focuses on the application of HTS in rechargeable batteries. In the negative electrode, a high-capacity composite negative electrode with Si nanoparticles was synthesized by high temperature thermal radiation. Si nanoparticles produced uniformly distributed nanoparticles in conductive RGO matrix (high-quality load > 2.5 mg cm^{-2}) without agglomeration. SiBs anode carbon material was prepared from switchgrass by HTS. The excellent sodium storage properties of this carbon material, combined with sustainable biomass sources and scalable synthesis methods, make switchgrass derived carbon an attractive SIBs anode material. In terms of current collector, a highly conductive RGO film with a DC conductivity of 3112 S/cm was prepared by HTS. Highly conductive RGO films are manufactured by a process of solution-based filtration followed by thermal reduction (773 K) and then high temperature reduction (joule heating 2750 K). The unique high-temperature reduction of joule heating can remove defects in GO effectively, improve the crystal structure of individual RGO nanosheets and densification between them, thereby increasing DC conductivity by 231 times, etc.

In summary, HTS technology has the characteristics of fast reaction, high efficiency, and low cost, and can be used to prepare a variety of nanomaterials. Multifunctional nanoparticles and high-quality carbon-based nanomaterials were synthesized by HTS, which are widely utilized in batteries and other fields. HTS will attract more and more research interest in the near future. Besides producing a variety of nanomaterials, HTS technology can be extended to a range of bulk/temperature sensitive materials, such as ceramics, organics, and polymers. The HTS approach facilitates the development of different battery systems (e.g., lithium/sodium/potassium ion batteries and lithium/sodium/potassium metal batteries) that have excellent electrochemical performance and thermal stability.

References

1. Yang C, Lv F, Zhang Y et al (2019) Confined Fe2VO4 subset of nitrogen-doped carbon nanowires with internal void space for high-rate and ultrastable potassium-ion storage. Adv Energy Mater 9(46):1902674
2. Dunn B, Kamath H, Tarascon J-M (2011) Electrical energy storage for the grid: a battery of choices. Science 334(6058):928–935
3. Nitta N, Wu F, Lee JT et al (2015) Li-ion battery materials: present and future. Mater Today 18(5):252–264
4. Nishi Y (2001) Lithium ion secondary batteries; past 10 years and the future. J Power Sources 100(1–2):101–106
5. Carmichael RS (2017) Practical handbook of physical properties of rocks and minerals. CRC Press
6. Kang KS, Meng YS, Breger J et al (2006) Electrodes with high power and high capacity for rechargeable lithium batteries. Science 311(5763):977–980
7. Zhao Y, Wang LP, Sougrati MT et al (2017) A review on design strategies for carbon based metal oxides and sulfides nanocomposites for high performance Li and Na ion battery anodes. Adv Energy Mater 7(9):1601424
8. Chen Y, Li Y, Wang Y et al (2016) Rapid, in situ synthesis of high capacity battery anodes through high temperature radiation-based thermal shock. Nano Lett 16(9):5553–5558

9. Bao W, Pickel AD, Zhang Q et al (2016) Nanocarbon paper: flexible, high temperature, planar lighting with large scale printable nanocarbon paper. Adv Mater 28(23):4566
10. Chen Y, Fu K, Zhu S et al (2016) Reduced graphene oxide films with ultrahigh conductivity as li-ion battery current collectors. Nano Lett 16(6):3616–3623
11. Yao Y, Fu KK, Yan C et al (2016) Three-dimensional printable high-temperature and high-rate heaters. ACS Nano 10(5):5272–5279
12. Nair RR, Blake P, Grigorenko AN et al (2008) Fine structure constant defines visual transparency of graphene. Science 320(5881):1308–1308
13. Li J, Dahn JR (2007) An in situ X-ray diffraction study of the reaction of Li with crystalline Si. J Electrochem Soc 154(3):A156–A161
14. Obrovac MN, Krause LJ (2007) Reversible cycling of crystalline silicon powder. J Electrochem Soc 154(2):A103–A108
15. Hatchard TD, Dahn JR (2004) In situ XRD and electrochemical study of the reaction of lithium with amorphous silicon. J Electrochem Soc 151(6):A838–A842
16. Netz A, Huggins RA, Weppner W (2003) The formation and properties of amorphous silicon as negative electrode reactant in lithium systems. J Power Sources 119:95–100
17. Kim H, Seo M, Park M-H et al (2010) A critical size of silicon nano-anodes for lithium rechargeable batteries. Angewandte Chemie-International Edition 49(12):2146–2149
18. Chan CK, Peng H, Liu G et al (2008) High-performance lithium battery anodes using silicon nanowires. Nat Nanotechnol 3(1):31–35
19. Ding J, Wang H, Li Z et al (2013) Carbon nanosheet frameworks derived from peat moss as high performance sodium ion battery anodes. ACS Nano 7(12):11004–11015
20. Li H, Shen F, Luo W et al (2016) Carbonized-leaf membrane with anisotropic surfaces for sodium-ion battery. ACS Appl Mat Inter 8(3):2204–2210
21. Wu L, Buchholz D, Vaalma C et al (2016) Apple-biowaste-derived hard carbon as a powerful anode material for Na-Ion batteries. Chem ElectroChem 3(2):292–298
22. Hong K-L, Qie L, Zeng R et al (2014) Biomass derived hard carbon used as a high performance anode material for sodium ion batteries. J Mat Chem A 2(32):12733–12738
23. David K, Ragauskas AJ (2010) Switchgrass as an energy crop for biofuel production: A review of its ligno-cellulosic chemical properties. Energy Environ Sci 3(9):1182–1190
24. Reddy N, Yang Y (2007) Natural cellulose fibers from switchgrass with tensile properties similar to cotton and linen. Biotechnol Bioeng 97(5):1021–1027
25. Zhang F, Yao Y, Wan J et al (2017) High temperature carbonized grass as a high performance sodium ion battery anode. Acs Appl Mat Inter 9(1):391–397
26. Lotfabad EM, Ding J, Cui K et al (2014) High-density sodium and lithium ion battery anodes from banana peels. ACS Nano 8(7):7115–7129
27. Johnson BA, White RE (1998) Characterization of commercially available lithium-ion batteries. J Power Sources 70(1):48–54
28. Myung S-T, Hitoshi Y, Sun Y-K (2011) Electrochemical behavior and passivation of current collectors in lithium-ion batteries. J Mater Chem 21(27):9891–9911
29. Braithwaite JW, Gonzales A, Nagasubramanian G et al (1999) Corrosion of lithiuim-ion battery current collectors. J Electrochem Soc 146(2):448–456
30. Lin X, Shen X, Zheng Q et al (2012) Fabrication of highly-aligned, conductive, and strong graphene papers using ultralarge graphene oxide sheets. ACS Nano 6(12):10708–10719
31. Zhao J, Pei S, Ren W et al (2010) Efficient preparation of large-area graphene oxide sheets for transparent conductive films. ACS Nano 4(9):5245–5252
32. Ferrari AC, Basko DM (2013) Raman spectroscopy as a versatile tool for studying the properties of graphene. Nat Nanotechnol 8(4):235–246
33. Dresselhaus MS, Jorio A, Hofmann M et al (2010) Perspectives on carbon nanotubes and graphene Raman spectroscopy. Nano Lett 10(3):751–758
34. Kim YD, Kim H, Cho Y et al (2015) Bright visible light emission from graphene. Nat Nanotechnol 10(8):676–681
35. Freitag M, Chiu H-Y, Steiner M et al (2010) Thermal infrared emission from biased graphene. Nat Nanotechnol 5(7):497–501

36. Liu S, Wang P, Liu C et al (2020) Nanomanufacturing of RGO-CNT hybrid film for flexible aqueous Al-Ion batteries. Small 16(37):2002856
37. Shen B, Zhai W, Zheng W (2014) Ultrathin flexible graphene film: an excellent thermal conducting material with efficient EMI shielding. Adv Func Mater 24(28):4542–4548
38. Worsley MA, Pham TT, Yan A et al (2014) Synthesis and characterization of highly crystalline graphene aerogels. ACS Nano 8(10):11013–11022
39. Xin G, Sun H, Hu T et al (2014) Large-area freestanding graphene paper for superior thermal management. Adv Mater 26(26):4521–4526
40. Xu Z, Liu Y, Zhao X et al (2016) Ultrastiff and strong graphene fibers via full-scale synergetic defect engineering. Adv Mater 28(30):6449–6456
41. Malard LM, Pimenta MA, Dresselhaus G et al (2009) Raman spectroscopy in graphene. Phys Reports-Review Sect Phys Lett 473(5–6):51–87
42. Ping W, Wang C, Wang R, et al (2020) Printable, high-performance solid-state electrolyte films. Sci Adv 6(47):eabc8641
43. Manthiram A, Yu X, Wang S (2017) Lithium battery chemistries enabled by solid-state electrolytes. Nat Rev Mater 2(4):1–16
44. Bitzer M, Van Gestel T, Uhlenbruck S et al (2016) Sol-gel synthesis of thin solid Li7La3Zr2O12 electrolyte films for Li-ion batteries. Thin Solid Films 615:128–134
45. Teng S, Tan J, Tiwari A (2014) Recent developments in garnet based solid state electrolytes for thin film batteries. Current Opinion Solid State Mat Sci 18(1):29–38
46. Cheng X-B, Zhang R, Zhao C-Z et al (2017) Toward safe lithium metal anode in rechargeable batteries: a review. Chem Rev 117(15):10403–10473
47. Wolfenstine J, Allen JL, Read J et al (2013) Chemical stability of cubic $Li_7La_3Zr_2O_{12}$ with molten lithium at elevated temperature. J Mater Sci 48(17):5846–5851
48. Tsvetkov N, Lu Q, Yildiz B (2015) Improved electrochemical stability at the surface of La0.8Sr0.2CoO3 achieved by surface chemical modification. Faraday Dis 182:257–269
49. Tsvetkov N, Lu Q, Sun L et al (2016) Improved chemical and electrochemical stability of perovskite oxides with less reducible cations at the surface. Nat Mater 15(9):1010–1016
50. Porz L, Swamy T, Sheldon BW et al (2017) Mechanism of lithium metal penetration through inorganic solid electrolytes. Adv Energy Mater 7(20):1701003
51. Aguesse F, Manalastas W, Buannic L et al (2017) Investigating the dendritic growth during full cell cycling of Garnet electrolyte in direct contact with Li metal. ACS Appl Mat Inter 9(4):3808–3816
52. Cheng L, Crumlin EJ, Chen W et al (2014) The origin of high electrolyte-electrode interfacial resistances in lithium cells containing garnet type solid electrolytes. Phys Chem Chem Phys 16(34):18294–18300
53. Wang C, Xie H, Ping W et al (2019) A general, highly efficient, high temperature thermal pulse toward high performance solid state electrolyte. Energy Stor Mat 17:234–241
54. Rao RP, Gu W, Sharma N et al (2015) In Situ neutron diffraction monitoring of $Li_7La_3Zr_2O_{12}$ formation: toward a rational synthesis of Garnet solid electrolytes. Chem Mater 27(8):2903–2910
55. Wang C, Wang Y, Yao Y et al (2016) A solution-processed high-temperature, flexible, thin-film actuator. Adv Mater 28(39):8618–8624
56. Qiao Y, Yi J, Guo S et al (2018) Li2CO3-free Li-O2/CO2 battery with peroxide discharge product. Energy Environ Sci 11(5):1211–1217
57. Chung DY, Jun SW, Yoon G et al (2015) Highly durable and active PtFe nanocatalyst for electrochemical oxygen reduction reaction. J Am Chem Soc 137(49):15478–15485
58. Qiao Y, Xu S, Liu Y et al (2019) Transient, in situ synthesis of ultrafine ruthenium nanoparticles for a high-rate Li-CO_2 battery. Energy Environ Sci 12(3):1100–1107
59. Li T, Pickel AD, Yao Y et al (2018) Thermoelectric properties and performance of flexible reduced graphene oxide films up to 3,000 K. Nat Energy 3(2):148–156
60. Yao Y, Huang Z, Xie P et al (2018) Carbothermal shock synthesis of high-entropy-alloy nanoparticles. Science 359(6383):1489–1494

61. Xu S, Das SK, Archer LA (2013) The Li-CO_2 battery: a novel method for CO_2 capture and utilization. RSC Adv 3(18):6656–6660
62. Liu Y, Wang R, Lyu Y et al (2014) Rechargeable Li/CO_2-O_2 (2:1) battery and Li/CO_2 battery. Energy Environ Sci 7(2):677–681
63. Qiao Y, Liu Y, Chen C et al (2018) 3D-printed graphene oxide framework with thermal shock synthesized nanoparticles for Li-CO_2 Batteries. Adv Func Mater 28(51):1805899

Chapter 4
Application of HTS in Carbon Materials

Contents

4.1 Introduction

Carbon materials have been broadly utilized in machinery, energy, biology, and other fields these years. At present, the preparation methods of carbon materials mainly include ball milling, ultrasonic stripping, solvothermal reaction, and chemical vapor deposition. These preparation methods have the disadvantages of high cost and low efficiency, and the prepared carbon materials are prone to surface oxidation and agglomeration, which reduce the properties of the materials. As a new synthesis method, HTS has many advantages, such as fast, low cost, universal, scalable, controllable, and environment-friendly. It has been widely utilized in the preparation of carbon materials, the composite of carbon materials, and the adjustment of the internal structure of materials.

Carbon materials (e.g., graphene films, graphene fibers, and graphene powders) can be prepared by the method of HTS. It can also be used to adjust the internal structure of carbon materials and prepare carbon materials into composite materials with other materials (metals, ceramics, and polymers). In addition, HTS technology

Y. Chen, *High Temperature Shock Technology*, Nanostructure Science and Technology,
https://doi.org/10.1007/978-981-19-8124-1_4

can also be used to prepare materials with high electrical conductivity and good comprehensive mechanical properties.

The application of HTS technology in carbon materials will be introduced below, mainly including its reaction device, preparation process, morphology characterization, and properties change.

4.2 Preparation of Carbon Materials by HTS

Graphene has excellent optical, electrical, and mechanical properties and has broad application prospects in materials science, energy, biomedicine, and other fields. Graphene can be prepared by top-down mechanical stripping [1] and bottom-up chemical vapor deposition (CVD) [2]. However, the output of mechanical stripping is pretty low and graphene synthesized by CVD is expensive. Therefore, high-efficiency and low-cost HTS method is used to prepare carbon materials such as graphene. Compared with traditional methods, carbon materials prepared by HTS have better properties, which provides a new idea for the preparation of carbon materials.

4.2.1 Preparation of Graphene Films by HTS

(1) A reduced graphene oxide film with high conductivity and mobility

Graphene oxide (GO) is a molecular layer of graphite with a variety of oxygen-containing functional groups. It is the most general chemically modified graphene and is the precursor of graphene synthesis. When GO is reduced by electrochemical, chemical, photochemical, or thermal approaches, the resulting reduced graphene oxide (RGO) is similar in structure to original graphene. The reduction procedure is critical to the quality of the synthesized RGO, because compared with chemical reduction, thermal reduction will not introduce impurities, and the reduction degree of GO can be controlled by simply changing the temperature of reduction [3]. The RGO consists of ordered SP^2 structures, which have nanometer size and are scattered between highly disordered SP^3 structures with oxygen-containing functional groups [4]. Vacancies, residual oxygen-containing functional groups, and other defects in RGO films tend to disperse or capture charge carriers passing through the graphite SP^2 network, restricting the conductivity and mobility of the films. Previously reported values of conductivity and mobility of RGO thin film are less than 1500 S cm^{-1} and 5 cm^2 V^{-1} s^{-1}, respectively [5], which prevents its application in flexible electronics, photovoltaic, energy storage, and other aspects.

In order to improve the electrical properties of RGO films, Professor Wang et al. [6] adopted a stable 3000 K joule heating approach. RGO films with electrical conductivity and mobility of 6300 S cm^{-1} and 320 cm^2 V^{-1} S^{-1} at room temperature were prepared by two-step reduction and bending of RGO films. The charge transport performances of RGO films reduced at different temperatures were studied. The RGO films reduced at 3000 K have a denser structure with fewer defects/impurities and a bigger graphite SP^2 structure, thus achieving its record performance.

GO dispersions were prepared by the improved Hummer approach [7], and independent GO films with controllable thickness were prepared by vacuum filtration. Firstly, the GO film was thermally reduced at 1000 K in an argon-filled furnace for 1 h. At a heating speed of 1 K min^{-1} from room temperature to 1000 K, the gas escape rate produced by oxygen-containing functional groups is small, which is conducive to maintaining the compact structure of RGO films. Then the RGO film reduced at 1000 K was suspended in a vacuum chamber above the substrate, and its two ends were bonded to the copper electrode by silver paste, as shown in the illustration of Fig. 4.1a. The RGO film has a curved shape that compensates for the structural changes caused by heat treatment. The joule heating process is carried out by utilizing direct current (DC) on the RGO film reduced at 1000 K, as presented in Fig. 4.1a. Firstly, the temperature of reduction is gradually increased to 2000 K, so that the remaining oxygen-containing functional groups in RGO film tardily escape. As the temperature of the electrode increases with the increase of the reduction temperature, the reduction temperature is lowered to reduce the temperature of the electrode in order to avoid the breakdown of the electrical connection caused by the loss of the silver slurry at high temperature. The temperature of the electrode is much lower than the temperature of reduction because the electrode is more efficient for the heat balance of the supporting substrate (heat radiation heating conduction) than the heat balance (heat radiation) of the RGO film. In the end, the temperature of reduction was increased fast to 3000 K to avert severe structural variations of RGO films. The overall success rate of this joule heating method is > 90%, which mainly depends on the sample size, particularly the width of the RGO film. The optical picture of the RGO film heated to 3000 K is illustrated in Fig. 4.1b, where the RGO film is bright white. The temperature is characterized by a fiber-coupled spectrometer that records emission spectra during joule heating (Fig. 4.1b).

Figure 4.1c manifests that the RGO film is oriented in the plane at random and its surface is smooth and successive. Figure 4.1d indicates the cross section of 3000 K reduced RGO film has many densely stacked RGO films that form lamellar structures. Compared with the original GO film, the RGO film reduced at 3000 K has a denser structure due to the removal of oxygen-containing functional groups attached to each graphene sheet.

The longitudinal resistivity and Hall resistivity of 3000 K reduced RGO film were measured by four-point probe measurement, as exhibited in Fig. 4.2. Figure 4.2a exhibits the relationship between temperature and longitudinal resistivity ρ_{xx} (T), where ρ_{xx} (T) increases with decreasing temperature. In order to enhance the conductivity of RGO, a lot of work has been done [6]. Chen et al. [8] found that the conductivity of GO film treated at 773 K was about 59 S cm^{-1}. Figure 4.2b shows the relationship between magnetic field and Hall resistivity ρ_{xy} (H), where ρ_{xy} (H) has a linear relationship with magnetic field. The negative slope manifests that the primary carriers in RGO films reduced at 3000 K are holes, which is in keeping with the fact that the residual oxygen-containing functional groups result in P-type doping. The

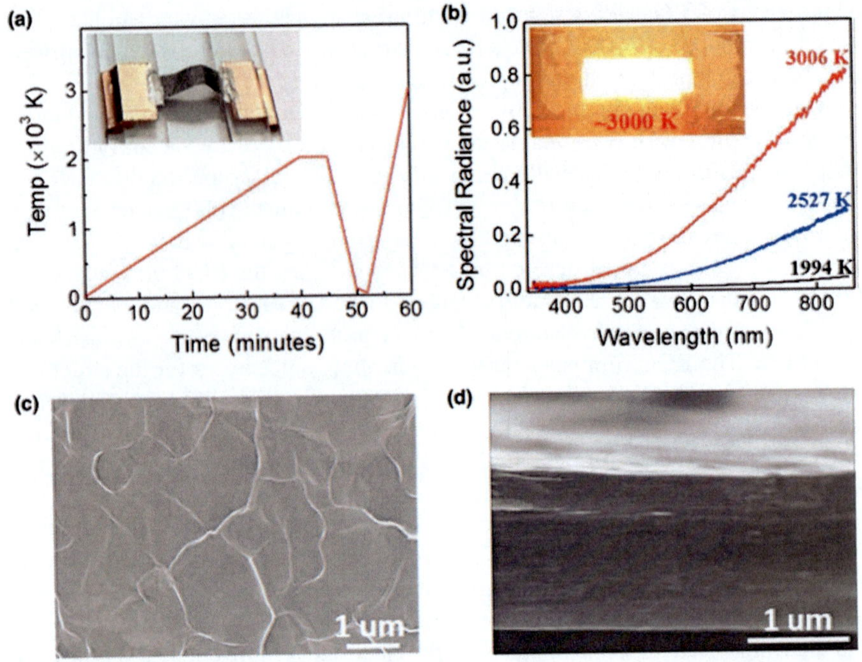

Fig. 4.1 Preparation of RGO film by joule heating **a** two-step reduction approach of joule heating by utilizing direct current through RGO film. The inset presents an experimental setup for a curved RGO film; **b** spectral radiation spectrometric measurements used to characterize the temperature of the heated RGO film. The inset presents an optical image of the RGO film heated to 3000 K; **c** SEM images of RGO films reduced at 3000 K; **d** SEM images in cross section

carrier mobility derived from Hall effect achieves 320 $cm^2\ V^{-1}\ s^{-1}$. Eda et al. [5] carried out tandem ammonia treatment and thermal annealing on the thin film at 200 °C, and the mobility obtained was merely 1 $cm^2\ V^{-1}\ S^{-1}$. However, due to the existence of residual oxygen functional groups and a little part of SP^2 structure, the mobility of previously reported RGO films is still very low [5]. The reason for the thermal reduction is the removal of the remaining oxygen-containing functional groups, which act as scattering centers and reduce the electrical conductivity and mobility of the RGO film. However, these oxygen-containing functional groups also act as doping sources that control the carrier density of RGO films [9]. In principle, a higher reduction temperature results in a smaller number of oxygen-containing functional groups and hence a lower carrier density.

(2) Thermoelectric properties of flexible RGO films below 3000 K

The development of ultra high temperature thermoelectric materials enables the thermoelectric performance of combustion power cycle to achieve new records and can also expand the scope of concentrated solar direct thermoelectric power generation.

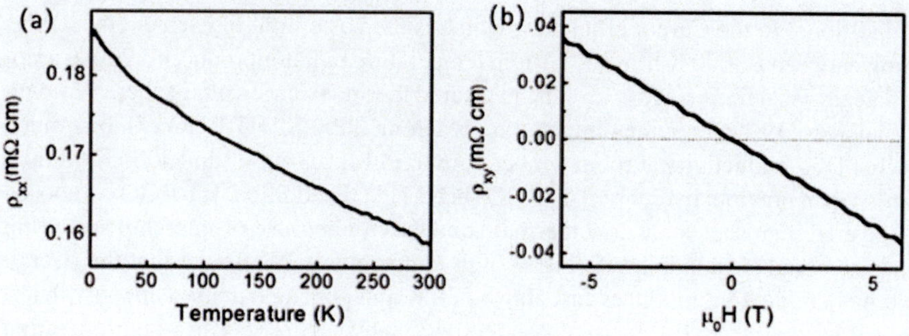

Fig. 4.2 Transport performances of 3000 K reduced RGO films **a** temperature dependence of the longitudinal resistivity ρ_{xx} (T) of RGO films at 3000 K; **b** magnetic field dependence of hall resistivity ρ_{xy} (H) of RGO films reduced at 3000 K

However, because of lacking appropriate materials, thermoelectric operating temperatures are limited to below 1500 K. To solve this problem, Li et al. [10] prepared a thermoelectric conversion material which was based on high-temperature RGO nanosheets, with performance up to 3000 K. After reduction treatment at 3300 K, the conductivity of the nanocrystals reaches 4000 S cm^{-1} at 3000 K, and the power factor is $S^2\sigma = 54.5\ \mu W\ cm^{-1}\ K^{-2}$. Measurements report that the thermoelectric properties of the film are up to 3000 K. The RGO film also has a high broadband radiation absorption rate and can be used as a radiation receiver and thermoelectric generator. This printable, lightweight, and flexible film is appealing to scalable manufacturing and system integration.

GO nanosheets were prepared by improved Hummer approach. Firstly, 1.5 g natural graphite nanosheets were added into 200 mL H_2SO_4/H_3PO_4 (volume ratio 9:1) mixed acid solution. Then, add 9 g of $KMnO_4$ to the solution slowly. The resulting suspension was heated in a 50 °C-water bath, stirred for 12 h, cooled to room temperature, and then poured into a mixture of 3 ml H_2O_2 and 200 ml ice. The mixture was washed once with 100 mL 30% HCl and a couple of times with distilled water. Independent GO films were obtained by filtration of dilute GO solution prepared by 0.65 μm pore size membrane (micropore). After drying, the GO film is separated from the film. The oxide film was then prereduced to RGO at a heating rate of 5 K/min in a 1000 K H_2/Ar (5%/95%) atmosphere in a tube furnace to ensure that the film was uniform before joule heating. The required amount of current is then applied through the film through joule heating, and the RGO film is treated at high temperatures (2000 K, 3000 K, and 3300 K). The emission spectrum is monitored utilizing a self-made experimental device as exhibited in Fig. 4.4a.

Using RGO bands as radiative heat sources, we tested the TE (thermoelectricity) HT-RGO receiver (Fig. 4.3a) over a wide temperature range (room temperature to 3000 K), which can operate at ultra high temperatures (up to 3000 K). The hot side of the film can reach 3000 K, while the cold side stays close to room temperature. The special high temperature performances of HT-RGO films have advantages in thermoelectric properties. The thermoelectric energy conversion efficiency is linearly proportional to the Carnot efficiency, which is the basic limit of each engine. Operating between a 3000 K film and a room temperature radiator produces a 90% Carnot efficiency (η_C) limit (Fig. 4.3c). The measured thermoelectric voltage increases with the increase of heat source temperature, reaching 3000 K. HT-RGO exhibits much higher DC conductivity at room temperature ($\sigma = 1450$ S cm^{-1}) and 3000 K (4000 S cm^{-1}) than previously reported on RGO sheets [11]. In addition, HT-RGO composed of layered nanosheets has low thermal conductivity because of interchip scattering and increased Umklapp scattering at high temperatures. We found that the overall thermal conductivity reduces fast above 530 K and conclude that it is mainly determined by the electronic thermal conductivity above 1600 K. Ultra-high operating temperature capability of 3000 K is also desirable for a wider operating range to prevent material degradation due to huge temperature fluctuations.

The optical and electrical performances of high temperature RGO films were studied utilizing a self-made experimental device (Fig. 4.4a). The temperature was determined by fitting the capture spectrum of irradiated RGO (Fig. 4.4b) to the Planck function of assumed gray emissivity ε_{grey}.

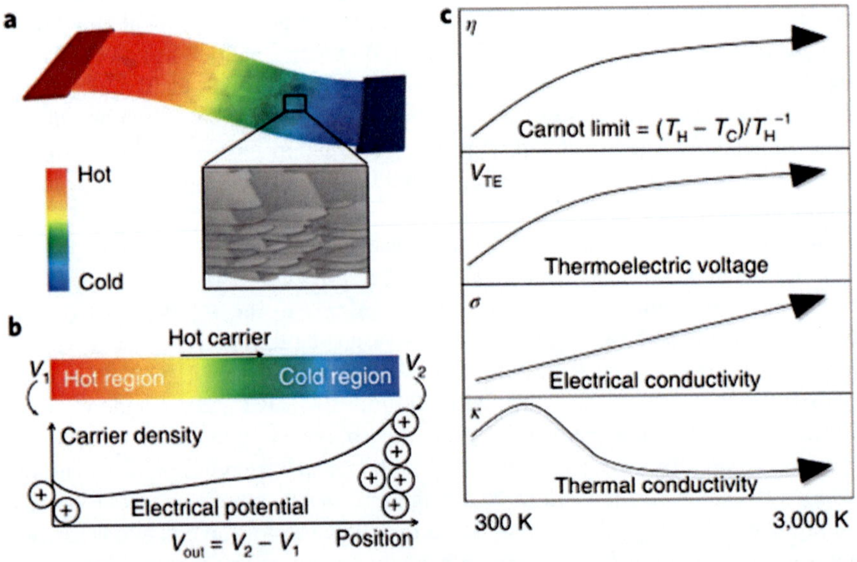

Fig. 4.3 HT-RGO running at high temperature **a** thermoelectric energy conversion devices using HT-RGO flakes. The illustration is a schematic diagram of the lamellar RGO thin sheet structure; **b** holes in the P-type RGO accumulate on the cold side due to asymmetric heating, which results in thermoelectric voltage. Hot: reach 3000 K, cold: room temperature; **c** for 3000 K on the hot side (cold side at room temperature), Carnot efficiency increases to 90%. The measured thermoelectric voltage and conductivity both increase with temperature, the latter reaching 4000 S cm^{-1} at 3000 K operating temperature

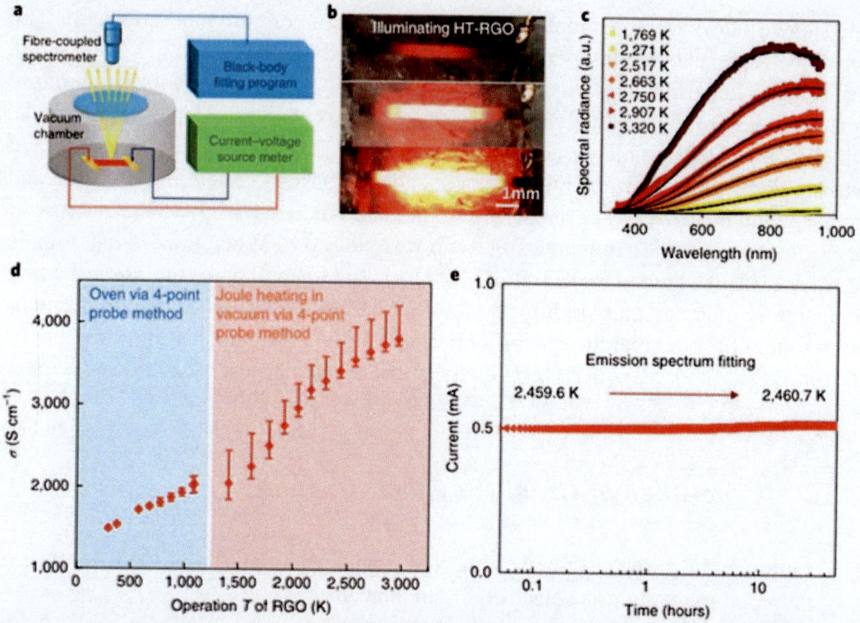

Fig. 4.4 High-temperature operation capability of HT-RGO film **a** experimental apparatus for in situ monitoring incandescent emission from self-supported HT-RGO films. Fiber-coupled spectrometers are utilized to analyze emitted radiation; **b** photographs of HT-RGO films heated by suspended joules at 3300 K; **c** the emission spectrum measured is taken from the center of the glass band; **d** relationship between electrical conductivity of RGO strip annealed at 3300 K and operating temperature; **e** RGO strips treated at 3300 K emit continuously for more than 50 h at a constant applied voltage of 6.3 V

$$I_\lambda(\lambda, T) = \gamma \varepsilon_{\text{grey}} \frac{2hc^2}{\lambda^5\left(\exp\left(\frac{hc}{\lambda k_B T}\right) - 1\right)} \tag{4.1}$$

where λ is wavelength, k_B, h, and c are Boltzmann constant, Planck constant, and the rate of light, respectively. The tunable parameter γ takes into account the uncharted geometric coupling factor (light collection efficiency) between the illuminated RGO film and the optical fiber input. As the input voltage increases, the RGO bands emit brighter over a greater part of their length. As shown in Fig. 4.4c, the emission spectra of RGO films measured at different temperatures agree well with Eq. (4.1). Stable high temperature operation up to 3300 K can be reached with external DC bias. Nevertheless, we noticed that RGO films cannot withstand temperatures greater than 3300 K, which may correspond to sublimation-induced failure.

The electrical conductivity σ of HT-RGO films annealed at 3300 K was further studied at temperatures reaching 3000 K. An obvious increase in conductivity with temperature makes a record-high conductivity of about 4000 S cm^{-1} possible compared to other RGO materials (Fig. 4.4d). The error line represents the mean and standard deviation of sample variation and film temperature inhomogeneity. The blue area represents the temperature range measured in a conventional oven, while the red area corresponds to the joule heating method. σ increases with temperature corresponding to thermal activation of charge carriers in lightly doped semi-metals, which predominate in increased phonon scattering. The conductivity of RGO band was studied by a four-probe measurement scheme during conventional heating and joule heating in an oven. The results are summarized in Fig. 4.4d. With the increase of operating temperature, an increasing trend was observed. When the driving voltage of 6.3 V is utilized to the 3300 K RGO film for more than 50 h, the emission spectral temperature only deviates slightly from 2459.6 to 2460.7 K (Fig. 4.4e). This demonstrates great high-temperature stability because of the highly crystalline structure of the film. A slight increase in DC conductivity can be owed to a self-healing procedure.

4.2.2 Preparation of Graphene Fibers by HTS

(1) Lightweight graphene fibers with high conductivity and high load flow were prepared by continuous electrothermal annealing

The recent increasing needs for miniaturized electronics call for the development of flexible, narrow, and lightweight channels with high current density as next-generation conductors. Though conventional metal conductors (such as copper and gold) have high conductivity and current-carrying capacity, their mass density is relatively high. In this respect, defect-free graphene fibers (GFs) are a good alternative due to their lightweight and high conductivity. Lee et al. [12] conducted electrothermal annealing of GFs at 3000 K for an extremely brief time (50 s) through successive current injection. This treatment resulted in carbonization and subsequent graphitization of GFs, which successfully repaired the structural disorder and crystallization defects of GFs. This successive procedure is beneficial to large-scale production of highly conductive graphite. The resulting meter-grade GFs showed a conductivity of 2721 S cm^{-1}, with a maximum ampacity of 3.84×10^4 A cm^{-2} and a specific current-carrying capacity of 4.67×10^4 A cm g^{-1}, higher than the corresponding value for commercial copper wires. This research offers a common and cost-effective technology for developing advanced fibers and thin films as lightweight conductors for related applications.

GFs is prepared as follows: GO is wet-spun to form graphene oxide fibers (GOFs), and a doping solution is prepared by dissolving GO (1.2wt.%) in deionized water without adding polymers or other components [13, 14]. A 1.2 wt.% GO solution was then placed into a syringe with a single nozzle and sprayed at a constant rate of 50 mm S^{-1} into a coagulating bath containing $CaCl_2$(5 wt.%) and NH_4OH (0.25 M) solutions. In the coagulation bath, the liquid crystal GO solution is transformed into gel fiber, and then into solidified GOFs to form GOFs [15]. The obtained GO was then dried and immersed in 50% hydroiodic acid solution at 353 K for 10 min to obtain chemically reduced GOFs (hereinafter referred to as CGFs). Wash the CGFs a couple of times to remove residual hydroiodic acid.

For electrothermal reduction, CGFs are suspended between two copper rod electrodes (gap: 3 cm). Two controllable motors were used to transport CGFs at a rolling rate of 1 mm S^{-1} between copper electrodes. The current is injected in an Ar environment. DC-regulated power supply device is utilized to input the target DC to the CGFs for joule heating during the rolling process. The time of target DC is controlled by programming software. Electrically treated CGFs are called EGFs.

Prior to the current treatment, the GOFs wound around the spool were reduced by high-intensity heat treatment to obtain GO with improved electrical conductivity. After high temperature treatment of GO, the defective carbon is transformed into SP^2-carbon structure, reaching the optimal conductivity of 145 S cm^{-1} [16]. In the absence of hydroiodic acid treatment, GOFs are electrically insulated and their resistivity is high. Hence, CGFs are suspended between two copper electrodes and subjected to joule heating of 15 mA current while rolling on two spools in the CCI procedure. As exhibited in the infrared thermal image in Fig. 4.5a, the temperature of the optical fiber reaches about 423 K. We found that applying an electric current to CGFs causes metastable volume expansion of the fibers when winding on two bobbin tubes. The side-view and cross-sectional SEM pictures of the fibers clearly show obvious morphological differences between the fibers before and after 160 mA joule heating (Fig. 4.5b–d) and (Fig. 4.5e–g). The diameter of CGFs (25 μm) increased nearly four times after the current treatment. CGFs (c-EGFs) treated with continuous current have a porous structure, which means that a mass of gas is generated in the period of electrically induced thermal annealing. When CGFs are heated at high temperatures, the deoxidation of GO and the decomposition of defective carbon structures are partially reduced, leading to the evolution of oxygen-containing and carbon-containing gases (e.g., CO and CO_2) [17]. Thus, as the spool continues to rotate, the new insertion portion of CGFs undergoes a sudden applied current and undergoes deoxidation, thereby releasing carbon-containing gases. The generation of large pores is inevitable because the resulting gas pressure is much higher than the pressure required to separate graphene sheets [17].

As exhibited in Fig. 4.5h, the conductivity of CGFs (0.113 S) increased by about 20 times after current treatment. Despite the large volume expansion, the conductivity of c-EGFs remained nearly constant at 141 S cm^{-1}, comparable to that of untreated CGFs (145 S cm^{-1}). This is mainly because of the recovery of graphite structure in the porous cell walls of CGFs. However, as shown in Fig. 4.5i, an increase in structural irregularity of the fibers (e.g., the creation of holes) leads to a decrease in the mechanical properties of the fibers after the current treatment. After CCI, Young's modulus of CGFs (18.6 GPa) was obviously reduced to 0.25 GPa, indicating poor load transfer in the porous structure [13, 18].

As shown in Fig. 4.6a, the CCI process is improved to reduce the volume expansion of fibers. The procedure is as follows: (Step 1) fixed current treatment, (Step 2) one-way fiber movement in the process of spool rotation, (Step 3) second fixed current treatment. For fixed current treatment, direct current was injected from 0 to 160 mA at a speed of 3.2 mA s^{-1} for 50 s to avoid porosity and glass fiber breakage. Then

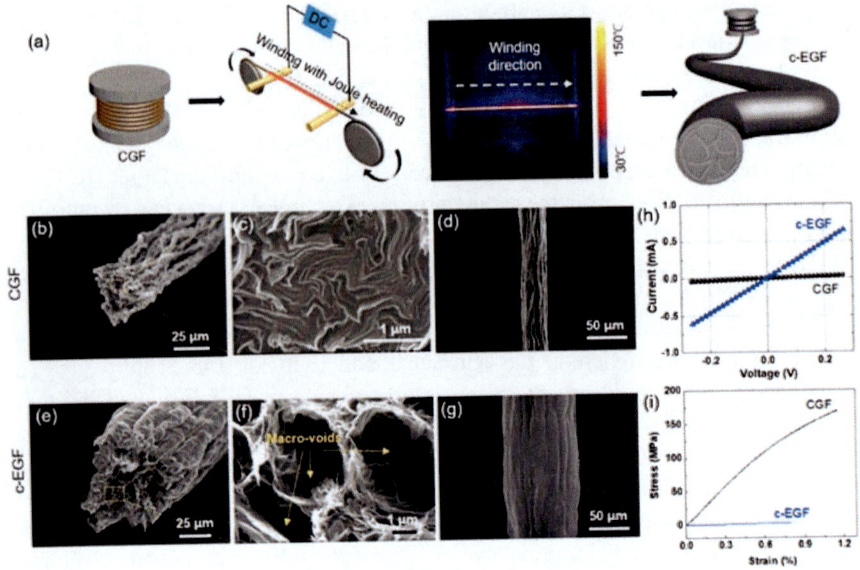

Fig. 4.5 **a** Continuous current injection (CCI) procedure diagram; CGF: **b** cross section, **c** magnified cross section, and **d** lateral scanning electron microscopy images; c-EGF: **e** cross section, **f** magnified cross section, and **g** side-looking SEM pictures; CGF and c-EGF: **h** I-V and **i** stress–strain curves

the EGFs are moved without current injection until the newly generated CGFs are transferred between the two copper electrodes. For thermal annealing, the input current subsequently increases from 0 to 160 mA at a speed of 3.2 mA s^{-1} for 50 s. Record the resultant voltage and corresponding resistance data relative to the processing time (Fig. 4.6b). In accordance with Ohm's law ($V = IR$), a steady voltage of about 65 V indicates a reduction in fiber resistance with a gradual increase in the input current, due to the structural changes and graphite recovery of the fibers caused by thermal annealing. Micron-grade fibers produced by S-CCI are wound around the bobbin, as shown in Fig. 4.6c. In addition, EGFs processed in a continuous manner (called s-EGFs) have mechanical flexibility and strength when processed (Fig. 4.6c, d). Cross section and side-view SEM pictures of s-EGFs are exhibited in Fig. 4.6e–g. Compared with c-EGFs (Fig. 4.5f), s-EGFs show a smaller pore size and a fiber diameter of about 25 μm. Dense, defect-healing glass fibers can be prepared by the S-CCI process without large voids and volume expansion.

Dense and defect-healing s-EGFs were obtained by efficient sequential current induction electrothermal treatment. This approach, based on the traditional CCI procedure, promotes mass manufacture of s-EGFs and systematic control of pore formation in the fiber internal structure. This procedure contributes to the defect repair of the graphite structure, resulting in the development of microcrystalline structure at the atomic and molecular level, and the enhancement of deoxidation and orientation of the graphene sheet. The s-EGF160 exhibits higher specific current

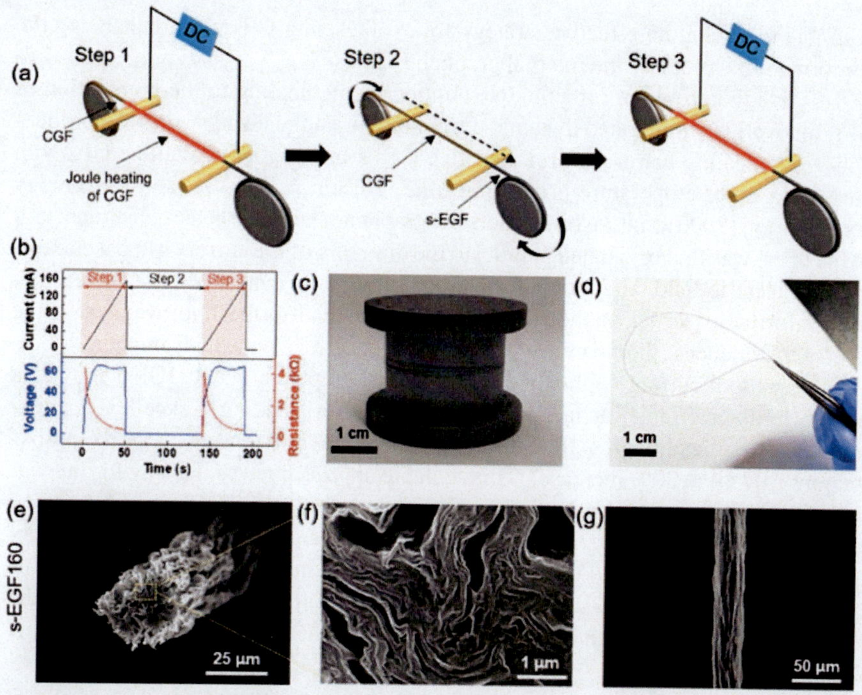

Fig. 4.6 **a** Schematic diagram of sequential continuous current injection (S-CCI) program; **b** changes in input current and corresponding GF voltage and resistance with respect to current input time during continuous joule heating; **c** photograph of the s-EGF160 wound around the spool; **d** photographs of s-EGF160, exhibiting the flexibility of the fibers; s-EGF: **e** cross section, **f** magnified cross section, and **g** lateral SEM images

densities than metals and a variety of carbon nanotubes and carbon-based fibers, as well as high conductivity after short electrothermal annealing processes. The s-EGF160 can be used as a new lightweight conductor in miniaturized circuits and high current-carrying conductors.

(2) Current alignment element during joule heating of graphene fiber

At present, the main synthesis method of graphene fibers is GO wet spinning, followed by GOFs reduction to remove oxygen-containing functional groups and heal defects. The reduction procedure can improve the electrical, mechanical, and thermal performances of fibers by several to tens of times. Among them, high temperature heat treatment is one of the most effective ways to reduce GOFs, remove non-carbon impurities, and enhance the crystallinity of GFs. However, this method is used at high temperatures (>2000 K) often needs a fairly long duration (>12 h), which leads to lower productivity and higher manufacturing costs. Besides, the conformational order of graphene sheets in fibers is an important parameter of fiber properties. Higher lamellar orientation can decrease structural disorder and enhance the compactness of GFs, resulting in better electrical and mechanical properties. Hence, it is highly desirable to exploit an effective strategy for synthesizing GFs while enhancing the conformational order of the internal graphene sheets.

To solve the problem, Liu et al. [19] applied joule heating to the preparation of GFs, innovatively integrated dynamic design into it, and realized successive production. Joule heating can transform defective GOFs into highly crystalline GFs with ultra-short high temperature processing time (20 min at 2000 K) and low energy consumption (2000 kJ m^{-1}). In addition, the graphene sheet (the basic constituent unit of the fiber) was first orientated parallel to the direction of the current using a current-induced electric field, and its conformational order was increased. The increase of the conformation order of the sheet is more beneficial to the improvement of the fiber performances. Compared to conventional thermal annealed graphene fibers (TGFs) without current application, joule-heated graphene fibers (JGFs) showed a higher Hermann orientation factor (0.73, 16% increase), leading to excellent conductivity (approximately 5.9×10^5 S m^{-1}, 11% increase) and tensile strength (approximately 1.07 GPa, 20% increase). This scalable approach paves the way for innovation in the rapid, continuous, efficient, and high-volume production of GFs, which is conducive to practical applications in wearable electronics, power cables, electromagnetic shielding, and other fields. Besides, the current-induced effect in this procedure can be utilized to the conformation control of component units in the assembled materials, which has great potential to further improve their macroscopic properties.

Dynamic joule heating (DJH) facility for successive manufacture of graphene fibers contains four main sections (Fig. 4.7a), a joule heating module, a coil-to-coil device, a vacuum supply system, and a gas supply system. During the DJH process, a beam of chemical RGOFs wound around two winding rolls passes through the joule heating zone at a programmed speed. The RGOFs wound on the feed roll are continuously moved to the receiving roll through the joule heating zone. At the same time, the temperature of JGFs was monitored by infrared camera. When an input current flows through the graphene fiber, the temperature of the fiber between the two electrode pulleys can rise to a stable value in a couple of seconds because of the joule heat effect. After joule heat treatment, the JGFs prepared show uniform gray contrast and high flexibility (Fig. 4.7b). High temperature in DJH procedure can efficiently remove carbon impurities and improve fiber crystallinity. JGFs were annealed at 1100 °C for 10 min (denoted as JGFs-1100) and then treated at 1100 °C for 10 min (denoted as JGFs-1500 and JGFs-2000). The removal of impurities was characterized by XPS, which revealed that the O 1 s peak at ≈533 eV decreased gradually with increasing temperature until it eventually disappeared at 2000 °C (Fig. 4.7c). Compared with RGOFs, JGFs (JGFs-2000) treated at 2000 °C showed negligible D-peak (Fig. 4.7d) and uniform 2D peak intensity (Fig. 4.7e) in Raman spectrum, indicating that the prepared JGFs had high quality.

As the temperature increases, the electrical and mechanical properties of JGFs improve simultaneously, with diameter shrinkage, thanks to improved crystal quality and sheet arrangement (Fig. 4.8a–c). Specifically, JGFs-2000 can realize excellent and uniform conductivity of $(5.9 \pm 0.2) \times 10^5$ S m^{-1}, which exceeds most polyacrylonitrile or bituminous carbon fibers. The tensile strength and Young's modulus of

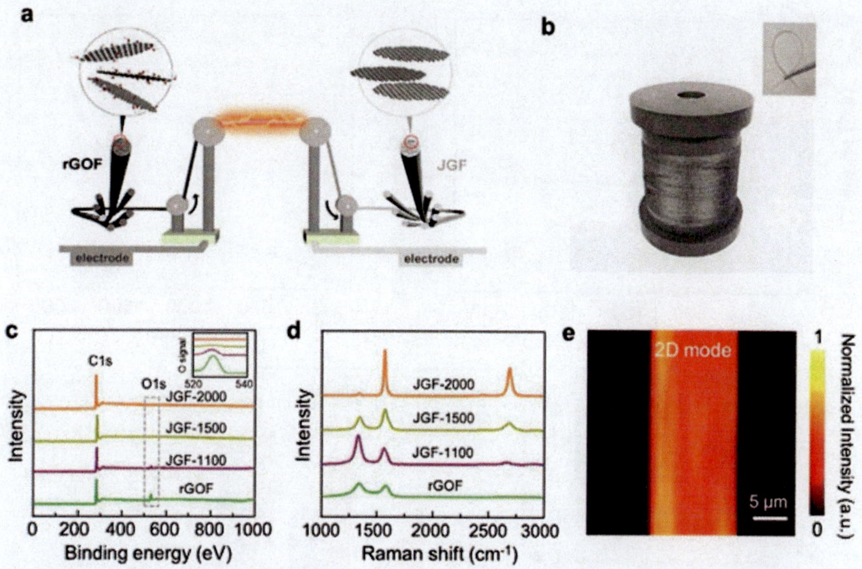

Fig. 4.7 Preparation of graphene fibers by DJH approach **a** schematic diagram of DJH system; **b** a roll of 100 silk JGFs. The illustration exhibits the flexibility of manufacturing JGFs; **c** the XPS spectrum of RGOFs. Inset is an enlarged O 1 s XPS spectrum, corresponding to the area in the gray dotted box; **d** Raman spectra of RGOFs and JGFs undergoing the same treatment as in **c**; **e** 2D Raman strength drawing of a single JGF-2000

JGFs-2000 are up to 1.07 ± 0.08 GPa and 116 ± 10 GPa, respectively (Fig. 4.8c). Compared to conventional heat treatment, the overall high temperature duration of graphene fibers during DJH is as low as 20 min, which is more than one order of magnitude shorter than conventional heat treatment, as shown in Fig. 4.8d.

The innovative DJH approach described above enables ultra-fast and continuous manufacturing of GFs. High temperature treatment durations of GFs in joule heating can be an order of magnitude shorter than conventional heat treatments earlier reported. During joule heating, the current flowing through JGFs can regulate the conformation of the graphene sheet component units so that they align along the fiber axis, resulting in a significant improvement in Hermann orientation factor, conductivity, and tensile strength relative to TGFs. The DJH approach with current-induced effect can be extended to successive synthesis of multipurpose assembly materials, realizing conformation control of component elements, and engineering their macroscopic properties.

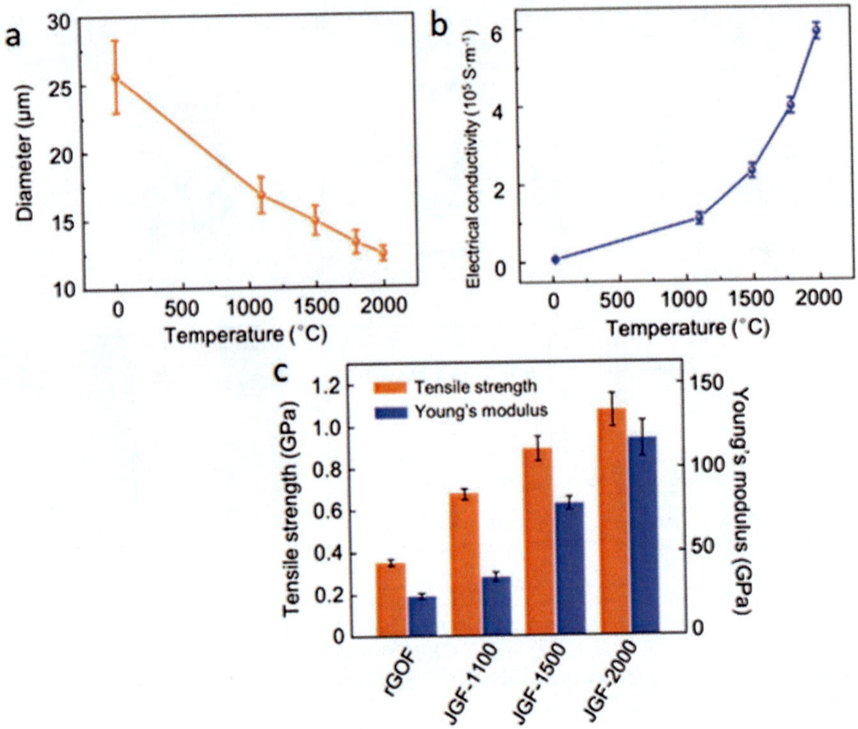

Fig. 4.8 Temperature-dependent structure and properties of JGFs **a** temperature-dependent diameter; **b** temperature-dependent conductivity; **c** temperature-dependent tensile strength and Young's modulus

4.2.3 Preparation of Graphene Powder by HTS

(1) Preparation of flash graphene by flash joule heating

Flash joule heating (FJH) can convert nearly any carbon-based precursor into large amounts of graphene. James Tour et al. [20] used FJH method to prepare flash graphene (FG) from carbon black and analyzed its morphology and properties. The results show that FG is composed of FG slices of turbine layer (tFG), and there is rotation mismatch between adjacent layers. The rest of FG is folded sheets of graphene, similar to non-graphitized carbon. To produce high-quality tFG sheets, FJH duration is 30–100 ms. Over 100 ms, the turbine laminates have time to generate AB stack and form massive graphite. The tFG is easy to strip by shear, so the FJH procedure has the potential to mass produce tFG without the use of chemicals or high-energy mechanical shearing for prestripping.

In order to synthesize FG, carbon black (CB) is encapsulated in a quartz tube and pressed between two copper electrodes. The copper cotton physically contacts the carbon black, as exhibited in Fig. 4.9a. The electrodes are connected to a capacitor bank with an overall capacitance of 60 mF and charged to a voltage of 120 V. A schematic diagram of the FJH system is shown in Fig. 4.10. The capacitor bank generates FJH through a controlled discharge of CB, rapidly heating CB to 3000 K in tens of milliseconds, and then cooling it to room temperature in seconds. During discharge, CB is fast heated and graphitized to form FG, as shown in Fig. 4.9b. Figure 4.9c shows the current and voltage at CB for 100 ms FJH duration. It is worth noting that a peak current greater than 400 A passes through CB, resulting in a fast loss of 30 kW power. This high-power consumption is responsible for the fast FJH. During FJH, CB forms planar crystals of tFG, which are arranged in the direction of current, similar to charred logs (Fig. 4.9d). The carbon black starting material consists mainly of small particles, which are mainly amorphous carbons with small graphite domains. After 100 ms of FJH, tFG sheets were dispersed with small graphitized carbon particles throughout the sample, leading to a holistic bottom-up synthesis of graphene.

The morphology of samples prepared by FJH method was analyzed. The atomic structure of tFG was observed by HR-TEM. Figure 4.11a shows the tFG sheet. Slight rotation mismatches between the sheets (indicated by non-parallel inserted white dashed lines) result in moire shapes that show up as stripes in the image merely where the mismatched sheets overlap. The rotation mismatch between these sheets is 5.4°.

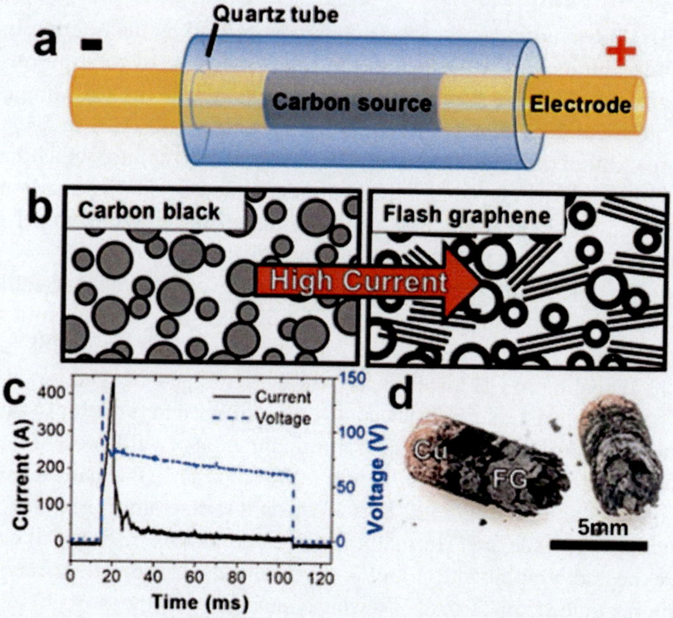

Fig. 4.9 General FJH features **a** schematic diagram of FJH facility; **b** schematic diagram of graphitization of carbon source after FJH. In the product (right), the circles mean graphitized carbon. These lines mean FG sheets; **c** the current and voltage through the carbon source during FJH is completed on a millisecond time scale; **d** FG photographs after FJH

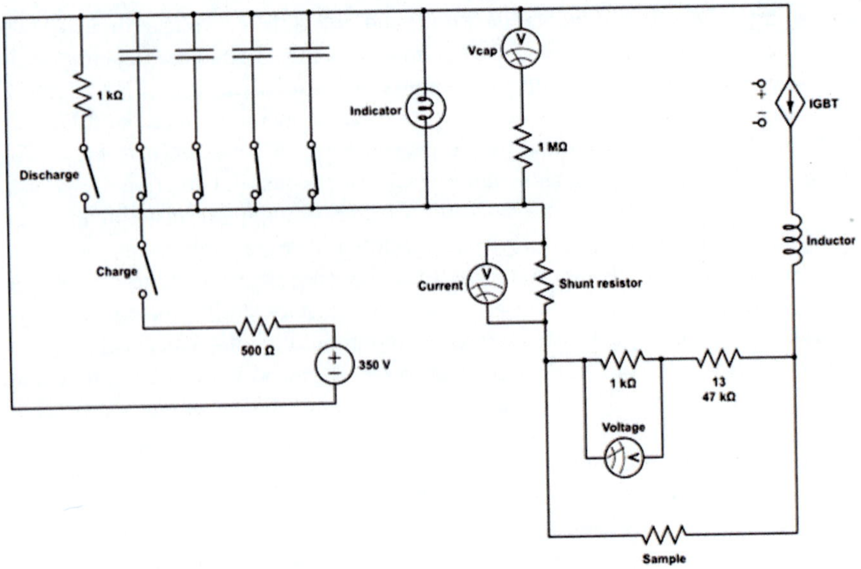

Fig. 4.10 Diagram of the FJH system

Figure 4.11b shows more clearly the moire shape caused by the rotation mismatch. Figure 4.11c exhibits similar stripes in the image because of rotation mismatches between adjacent graphene sheets. As shown in Fig. 4.11b, the FG sample is made up of tFG flakes and smaller graphite carbon particles. HR-TEM images of small particles are exhibited in Fig. 4.11d, e. The material is graphite, which has lots of bends and is usually 3 to 8 layers thick. It is similar to non-graphitized carbon and will be called folded graphene in this study. TEM images of the two forms of graphene show an atomic spacing of 0.34 nm.

The morphological differences between tFG and folded graphene indicate that the formation conditions are not uniform. Utilizing molecular dynamics simulation of AIREBO potential, we analyzed the procedure and results of rapid high temperature annealing of CB material, as exhibited in Fig. 4.12. The initial structure was designed to consist of 66% CB centroid particles and 34% loose amorphous carbon, surrounded by graphite walls to mimic the quartz walls present during FJH (Fig. 4.12a). After the initial pretreatment at 1000 K for 5 ms, the material was heated and subsequently annealed at 3500 K at a constant temperature (Fig. 4.12b), while graphitization (Fig. 4.12c) and Hermann orientation function (Fig. 4.12d) were monitored. As expected, graphitization levels, especially the carbon atom fraction in sp^2 configuration, ranged from 60 to 85% during annealing, with the most obvious change occurring during initial heating (Fig. 4.12c insets). It is interesting that the fluidity of the material within the centroid particles is obviously reduced compared with the loose amorphous carbon, leading to a small increase in graphitization because of the slow annealing of the defects in the graphene centroid shell. In the meanwhile, the surrounding amorphous carbon has no rigid preexisting structure and undergoes

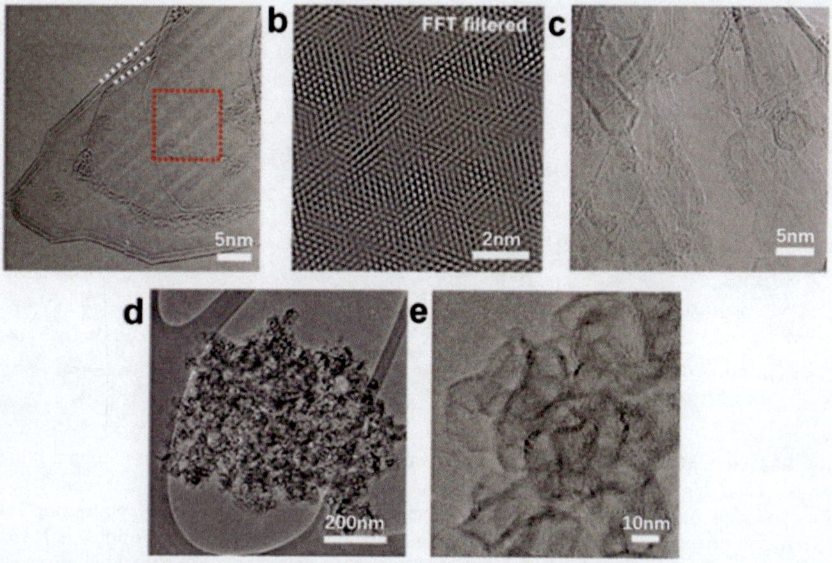

Fig. 4.11 Microanalysis of FJH products **a** TEM image of tFG sheet; **b** FFT filtered image of the inserted region in the image showing moire pattern; **c** TEM image of tFG shows more stripes, manifesting rotation mismatch; **d** TEM images of folded graphene produced by the FJH process; **e** HR-TEM images of folded graphene

fast graphitization, leading to a graphene-like coating covering the centroid particles (Fig. 4.12e–h).

Guided by experimental observations, the arrangement of graphite domains in the structure during annealing has been analyzed using the Hermann orientation function θ, which is generally utilized in polymer science. Calculate $\theta = (3\langle \cos 2\alpha \rangle - 1)/2$, where α is the angle between the normal of the partial direction of the graphite plane and the characteristic direction selected for horizontal (Fig. 4.13); direction functions shall be equal to 1, − 0.5, and 0, randomly oriented domains aligned and perpendicular to the graphite wall. During the annealing process, no obvious arrangement changes were observed (Fig. 4.13d), and the values specified in the initial configuration were mainly retained. Even after long-term thermal annealing at high temperature (3500 K), the structural changes of tFG generated from FJH were not obvious, and only folded graphene was observed (Fig. 4.13h). From these observations, we can conclude that folded graphene forms in the region of the material that undergoes simple thermal annealing. Meanwhile, current and its directivity may guide the generation of tFG. Based on thermal simulation and experimental observation, we can predict the formation of two different FG forms. This difference results from the preexisting morphology of carbon black with characteristic centroid particles (consisted of concentric graphite shells) and surrounding amorphous carbons that do not show any overall morphology (Fig. 4.12e, f). Interestingly, there is significantly less variation in the fraction of SP^2 hybrid carbon in the centroidal particles compared

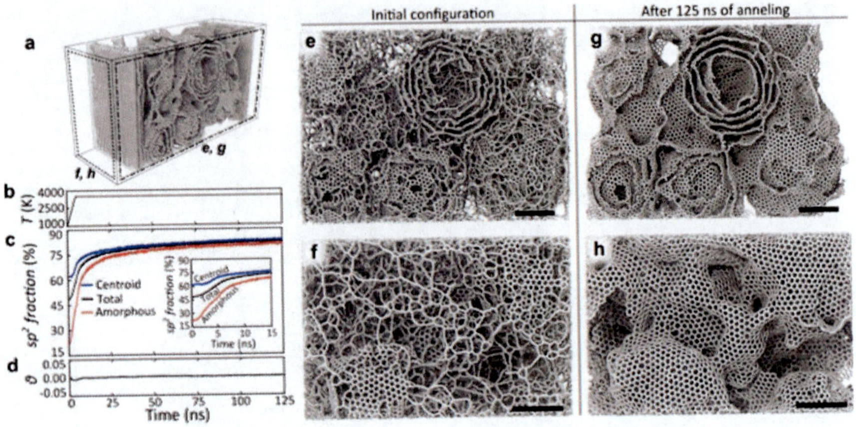

Fig. 4.12 Simulation of flash graphene generation **a** the structure of amorphous carbon is utilized for molecular dynamics simulation of high-temperature annealing. Vertical graphite walls face the front and back of the frame; **b** temperature distribution during annealing; **c** graphitization level; **d** Hermann orientation function; **e**, **f** initial configurations at two different positions; **g**, **h** visual comparison of the final configuration at two different positions, exhibiting the obvious level of graphitization after annealing

with the amorphous material (Fig. 4.12c), manifesting an increased deignification barrier for atoms in the preexisting graphene structure.

In addition, Loung et al. [21] used FJH approach to dispose of cheap carbon sources, e.g., biochar, coal, carbon black, rubber tires, waste food, petroleum coke, and mixed plastic waste to obtain gram-grade graphene in less than 1 s. The FG was characterized by morphology, and the FJH process was simulated. Wyss et al. [22] prepared ^{13}C turbine layered flash graphene using FJH method and characterized

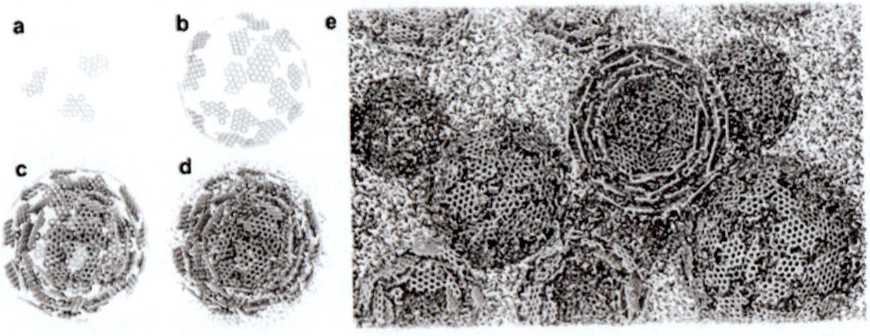

Fig. 4.13 Initial configuration of atom simulation from **a** single graphite flakes arranged in **b** spherical shells **c** centroid particles formed by adding single carbon atoms; **d**, **e** the final structure consists of eight centroid particles, graphite walls, and approximately 30% randomly positioned carbon atoms

the samples by solid-state X-ray photoelectron spectroscopy, IR, NMR, Raman, and inductively coupled plasma mass spectrometry.

The above proves that different carbon sources can be converted into turbine lamellar flash graphene in a very short time by FJH method, which provides a new idea for preparing graphene powder.

(2) Graphene is extracted from waste

PW pollution has become one of the most urgent environmental problems now. Most PW ends up in landfills and oceans, resulting in the generation of microbes and nanoplastic that threaten marine life, microbes, beneficial bacteria, and humans. Besides, plastics produced by petrochemicals have a high carbon footprint. Crude oil must be refined, distilled, refined, and purified to form fossil feedstock and then further processed into plastics in energy-intensive facilities that emit large amounts of greenhouse gases. Extra greenhouse gases are emitted during plastic molding and when it is shipped to customers. After this strong carbon footprint, most synthetic plastics are used only once before being dumped in landfills or ending up in ocean waterways. Thus, upgrading PW to higher value materials and chemicals is both environmentally and economically advantageous.

Algozeeb et al. [23] described an alternative method of chemical and physical recovery for PW treatment based on the DC flash joule heating (DC-FJH) approach (Fig. 4.14), which converts the carbon source into graphene, a process that forms flash graphene. The technique depends on electricity to induce FJH in a PW, which heats the carbon source to high temperatures in a brief time. This work exhibits that alternating current flash joule heating (AC-FJH) (Fig. 4.15) is advantageous over direct current (DC-FJH) when dealing with PW and it can last for several seconds, in this case 8 s. This process releases the necessary volatile compounds to produce intermediate AC flash graphene (AC-FG) with an I_{2D}/I_G peak ratio between 1.2 and 0.5, and a high-intensity D-band when characterized by Raman spectroscopy. Then, on a single DC-FJH pulse, the intermediate AC-FG is converted to a very high-quality turbine laminated FG (tFG) with an I_{2D}/I_G peak ratio between 1 and 6, and a low-intensity D-band when characterized by Raman spectroscopy. This sequential AC and DC (ACDC) flash procedure is efficient for cycling both single-flow thermoplastics and PW mixtures. Given graphene's high stability to typical microorganisms, chemical degradation, and thermal degradation, the technology provides a way to convert PW into a stable natural carbon form with low toxicity.

PW products mentioned in this study contain PET from carbonated beverage bottles, HDPE from milk cans, PVC from water pipes, LDPE from single-use plastic bags, PP from single-use straws and food packaging, and PS from single-use coffee cups. The PW is sanded or cut utilizing a cutting machine to obtain a powder size of 1–2 mm. The powder plastic was then mixed with 5 wt.% carbon black to get a conductive mixture (Fig. 4.16).

AC-FJH process: Powder is filled between two copper electrodes in a quartz tube (tube thickness: 2 mm, inside diameter: 8 mm, length: 5 cm). A sample with a resistance of 120 ~ 125 Ω was obtained by compressing 0.5 g of plastic. An alternating

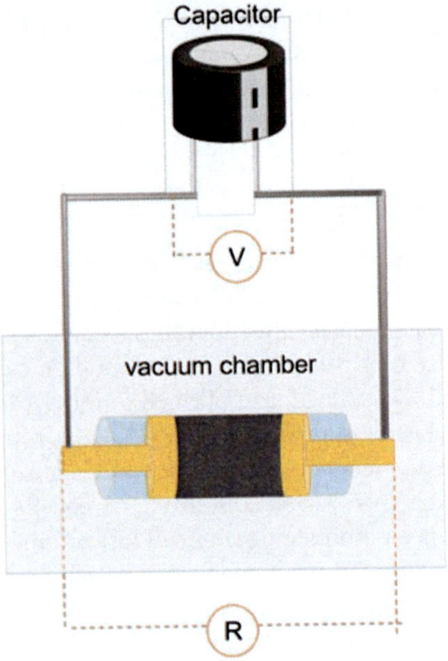

Fig. 4.14 Simplified solution of the DC-FJH device

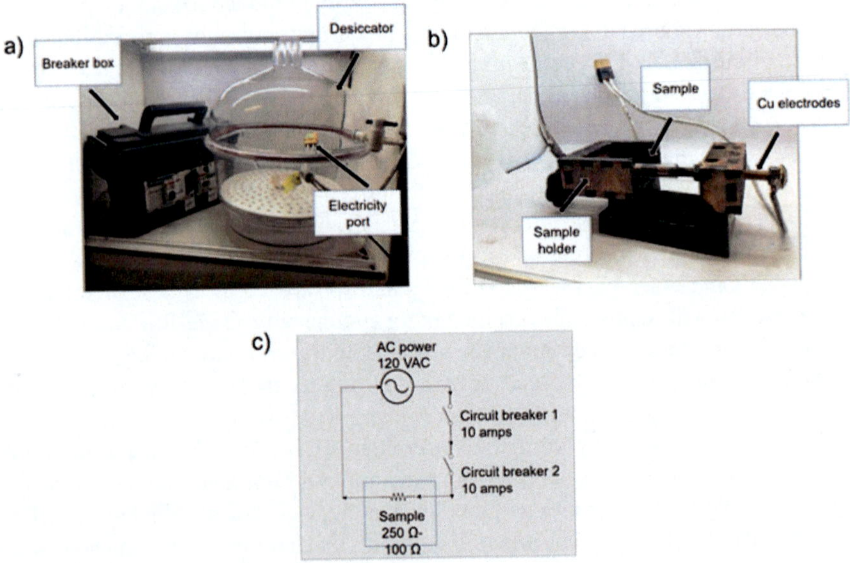

Fig. 4.15 Image of AC-FJH PW sample pretreatment equipment **a** the sample holder is housed in a plastic dryer to protect the operator. Electrical ports of 10 A circuit breaker box and dryer are indicated by label; **b** close-up view of sample holder, sample, and copper electrode; **c** circuit diagram of AC-FJH equipment

Fig. 4.16 Photographs of post-consumption plastic received from recyclers (left to right). Use commercial cutting machine after cutting. After mixing with 5 wt.% carbon black and use AC-FJH to further convert to FG

current (120 V, 60 Hz) is used to the sample for approximately 8 s in a vacuum dryer (approx. 10 mm Hg) to aid in venting. Figure 4.15 shows the AC system.

DC-FJH process: After AC-FJH, the sample is DC-FJH. A capacitor bank consisting of 10 450 V-and-60mF capacitors charges to 110 V and allows a discharge time of 500 ms to obtain a high-quality FG. Figure 4.14 shows the DC circuit.

The mass of FG was determined by Raman spectroscopy. It was found that AC-FJH formed FG with different I_{2D}/I_G peak ratio and different D-band strength. Figure 4.17a shows the average characteristic Raman spectra of FG obtained from the AC-FJH pretreatment procedure, exhibiting wide 2D and G bands and a large number of D bands. Utilizing a single 500 ms DC pulse (Fig. 4.14 for the DC circuit outline), the quality of AC-FG is obviously improved, thus obtaining high-quality turbine lamina FG (tFG) from a variety of plastics (Fig. 4.17b, the plastic mixture was 40% HDPE, 20% PP, 20% PET, 10% LDPE, 8% PS, and 2% PVC). The tFG obtained from AC-FJH and DC-FJH is called ACDC-tFG. Figure 4.17a, b shows that compared with AC-FG, the peak value ratio of I_G/I_D in ACDC-tFG is significantly increased. Figure 4.17c exhibits the Raman spectra of ACDC-tFG in PVC with I_{2D}/I_G peak ratio equal to 6. The observed TS_1 and TS_2 bands indicate the pure turbine layer morphology of tFG. Figure 4.17d exhibits the temperature profile of the AC-FJH process collected by a built-in infrared (IR) spectrometer. Blackbody radiation curve fitting was performed on the collected data, and it was found that the temperature rose to 2900 K during AC-FJH. DC-FJH shocks are known to reach 3100 K, which is the temperature needed to obtain high-quality graphene.

Using FJH to convert PW into high-value material using a small amount of energy. Large-scale use of FJH technology to treat PW has the potential to cut down greenhouse gas emissions from plastic preparation to recycling. However, to take full advantage of this approach, a full life cycle analysis is required. It is reported that 38.8 kJ of energy is required to produce 1 g of pure PET, while only 23 kJ is required to process PW using the FJH method, in order to cycle up to tFG and not just recycle. Graphene is considered to be a stable carbon source with an extremely elastic structure. Like graphite, graphene degrades at a slower rate of microbial degradation,

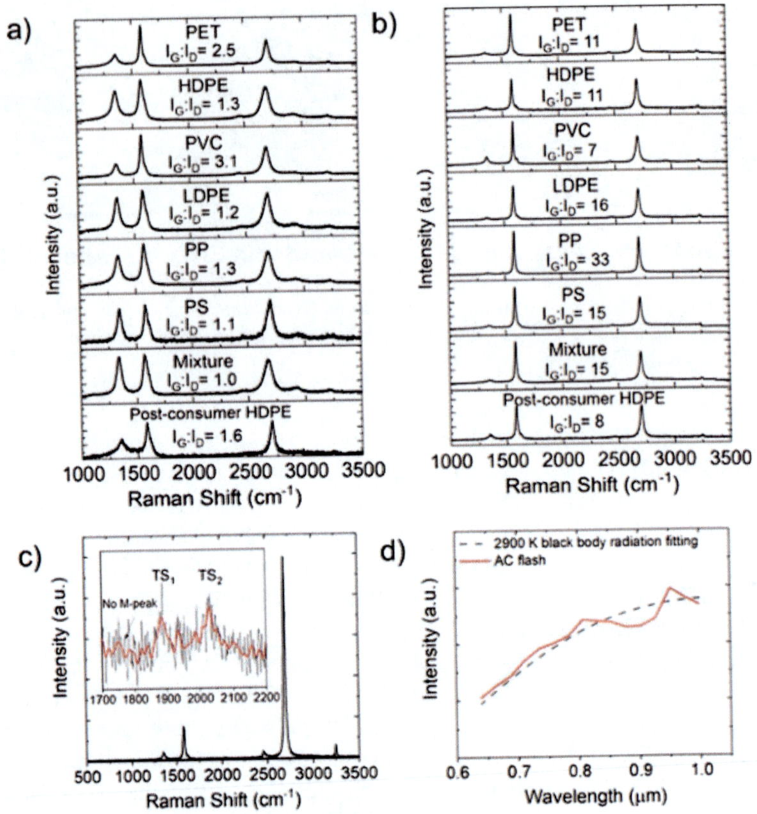

Fig. 4.17 Characteristic Raman spectra of **a** AC-FG and **b** ACDC-tFG; **c** Raman spectra of the high turbine layer FG of ACDC-tFG observed from PVC, exhibiting the temperature distribution of the turbine layer FG band in the extended spectrum; **d** utilizing an infrared spectrometer and blackbody radiation fitting to collect the AC-FJH process

slowing its reentry into the carbon cycle. Thus, the FJH of a PW should be viewed to loop up a PW.

In addition, Wyss et al. [24] adopted the method of FJH to transform PW into tFG. TFG can be utilized to improve the physical and mechanical performance of PVA without further functionalization or purification. Compared with pure PVA, the failure strain increases by 50% and water absorption decreases by 500%. The results show that recycling PW to high-value tFG and expanding its implementation and application by adding tFG-reinforced polymers provide an environmentally friendly way to reduce PW through economic incentives for chemical recovery. In addition, Advincula et al. [25] used the FJH method to produce high-quality tFG from rubber waste. The energy cost of manufacturing tFG with FJH is less than us $100 per ton of electricity for rubber waste, making FJH attractive for mass production of tFG and providing a good way to recycle harmful environmental pollutants. This will further lead to the wider use of tFG in different applications, leading to stronger and lighter materials at lower cost.

4.2.4 *Preparation of Other Carbon Materials by HTS*

(1) Ultra-fast controlled phase transition of flash joule heating

Flash joule heating (FJH) is a progressive material synthesis technology that has been utilized to produce high-quality carbon materials. Using FJH technology, the carbon-based material was successfully transformed into a large amount of turbine-layered graphene. But the generation of other carbon allotropes, for example, nanodiamond and concentric carbon materials, as well as covalency functionalization of different carbon allotropes by the FJH procedure, remains challenging. Chen et al. [26] reported solvent-free FJH synthesis of three different fluorinated carbon allotropes: fluorinated nanodiamonds, fluorinated turbine-layered graphene, and fluorinated concentric carbon. This is achieved by the millisecond heating of organofluoride compounds and fluoride precursors. Spectral analysis confirmed the change of electronic states and the presence of a variety of short- and long-range orders in different fluorocarbon allotropes. This experiment further proves that FJH time can control phase evolution and product composition.

Figure 4.18a exhibits the FJH settings. The different kinds of reactants are tightly packed inside PTFE tubes with tungsten carbide rods with graphite spacers. The reactant is a mixture of various organic fluorine compounds. Organic fluorine compounds have intrinsic SP^3 hybrid carbon atoms, which can reduce the formation energy of FND. Selective etching of carbides such as silicon carbide (with inherent SP^3 hybrid carbon atoms) has been reported to remove silicon atoms from the lattice of the carbide and induce direct formation of crystalline diamond structure carbon. Add carbon black (CB) or flash graphene (FG, 20wt.%) and mix evenly to obtain reactants with desired conductivity. During FJH, the large amount of energy generated by electric heating triggers the generation of a variety of fluorocarbon allotropes of different precursors. By using PTFE and CB (20wt.%) as reactants, microscopic imaging can be used to determine the morphology, thus distinguishing the formation of specific phases of carbon materials at different flash stages. Fluorinated amorphous carbon (FAC) and fluorinated nanodiamonds with high crystallization and polyhedral shape can be observed sequentially by controllable adjustment of the flash conditions. FND (SP^3-carbon)), fluorinated flash graphene (FFG, SP^2-carbon), and fluorinated concentric carbon (FCC), as shown in Fig. 4.18b–e.

HR-TEM was performed on the samples to determine the atomic structure of different carbon allotropes in Fig. 4.19. As exhibited in Fig. 4.19a, the irregular stripes reflect the amorphous nature of the initial product after a brief flash pulse. The inset exhibits the fast Fourier transform (FFT), which results in fuzzy rings, which also indicate amorphous properties. In contrast, FND (Fig. 4.19b–d) shows plane distances of 0.21 and 0.18 nm, which are the characteristic distances of {111}

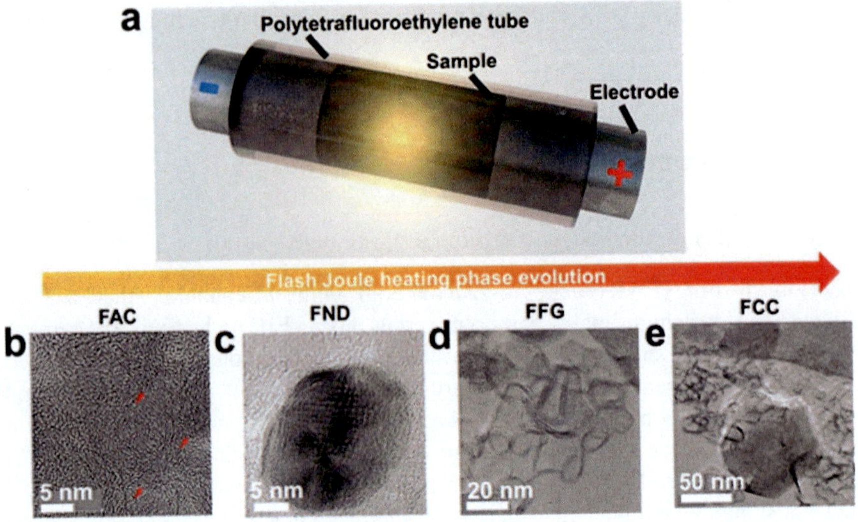

Fig. 4.18 Phase evolution of carbon materials **a** schematic diagram of the FJH facility for studying phase evolution; **b–e** HR-TEM images exhibit the specific phases of carbon materials at different flash stages; **b** FAC **c** FNDs, **d** FFG, **e** FCC products

and {200} diamond cubes, respectively. As can be seen from the HR-TEM image in Fig. 4.19b, FND terminates from the {111} plane, which is the lattice plane with the lowest energy compared with other low refractive index planes. The corresponding FFT image in Fig. 4.19c exhibits the Bragg spot with quadruple symmetry, which reflects the {111} plane. Figure 4.19d exhibits the distribution of FND. The contrast difference in TEM images enables the FND to be distinguished from the FAC substrate. Figure 4.19e shows TEM images of another typical carbon allotrope, FFG. The inset exhibits several different Bragg points of FFT, reflecting the misorientation between different graphene sheets in the stacking direction [002], and confirming the turbine stratification characteristics of FFG. The average size of FFG is 20 nm, which is like the non-functional FG derived from CB. As shown in Fig. 4.19f, longer heating time results in the generation of FCC with polyhedral structure and high crystallinity. HR-TEM images show that the graphene sheets with increasingly varying sizes share a common center, and that the periphery of the FCC has higher crystallinity than the central area. In the central area, amorphous areas and grain boundaries can be observed. The illustration in Fig. 4.19f exhibits the FFT image, which shows arc fragments with bright and distinct Bragg spots, reflecting the presence of graphene layers with different orientations along the C-axis, and the coexistence of AB stacked graphene and turbine-layered graphene.

Above by adjusting the flash conditions, in < 1 s different fluorinated carbon allotropes (e.g., FFG, FNDs, and FCC) were synthesized within 1 s. This method

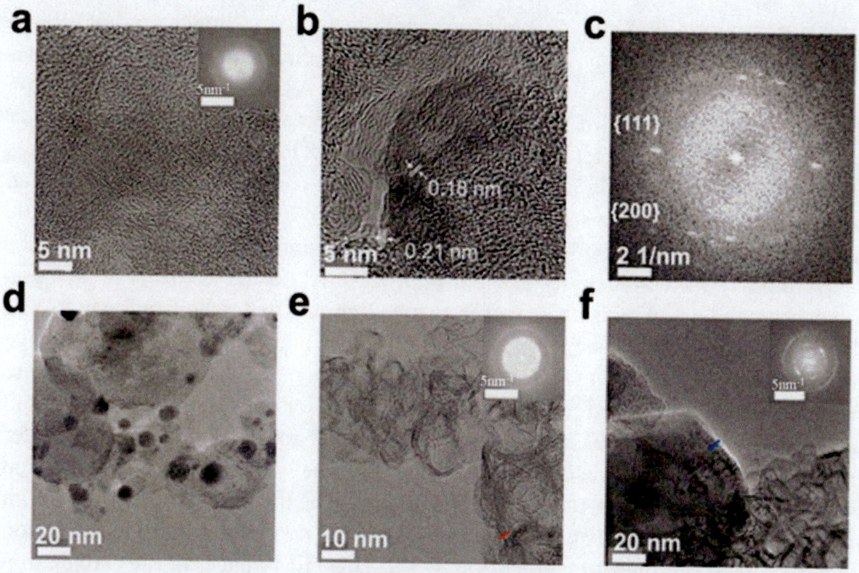

Fig. 4.19 Microstructure analysis of carbon material in different phase evolution stages **a** HR-TEM image of FAC. The illustration is showing the FFT of the blur ring; **b** HR-TEM images of FNDs; **c** FFT image of the chosen region in **b**; **d** TEM images showed the distribution of FNDs; **e** TEM image of FFG. The illustration is an FFT exhibiting multiple different Bragg spots; **f** TEM image of FCC with high crystallinity. The illustration is FFT, exhibits arc fragments with distinct Bragg spots

is more efficient, fast, and economical. This work further extends the FJH directional synthesis to produce various carbon allotropes except FG, thus broadening the application range of phase-selective synthesis using FJH process.

(2) Biomass is converted into highly conductive graphite carbon at ultra high temperature

Lignin based on polyphenol propane is the second most sufficient biopolymer on earth and is commonly considered as a renewable resource of carbon materials. Because of its high carbon content, lignin can be thermally converted into different carbon forms under inert conditions, for example, carbon fiber [27], activated carbon [28], porous carbon materials [29], and amorphous carbon [21, 30]. Nowadays, joule heating has been introduced as a simple and energy-saving method that can heat carbon-based materials to a temperature of 3000 K in just a few tens of milliseconds [31–33]. Advincula et al. [34] converted wood carbon into graphite carbon by joule heating, and the prepared graphite carbon had an ultra-high conductivity of 4500 S cm^{-1}.

During joule heating, both ends of the carbonized GO lignin film are connected to a copper electrode with a conductive silver slurry. The film is suspended on two separate slides to refrain from contacting with the bottom slide. Joule heating is executed in a protected atmosphere of argon. The mixed GO/lignin composite film manufactured by casting at 333 K was first thermally carbonized at 873 K to reduce the oxygen/carbon film (Fig. 4.20a). Figure 4.20b shows an image of RGO lignin carbon film heated with power of 26.5 W. The spectral emissivity of the joule-heated film in Fig. 4.20b is exhibited in Fig. 4.20c. As exhibited from the figure, the film temperature is 2465 K. The structure of lignin and carbon before and after joule heating was characterized by high-resolution TEM images. These pictures exhibit that the completely amorphous carbon in the carbide film (Fig. 4.20d) can be converted to highly crystalline and ordered carbon by a high-temperature joule heating procedure. The conductivity of as-cast GO lignin film is poor.

To improve the conductivity of the film, it was first heated to 873 K in an argon atmosphere to locally reduce GO and carbonize lignin. After carbonization at 873 K, the smooth film surface becomes granular with a large number of clumps in the range of tens to hundreds of nm (Fig. 4.21a, b). We hypothesize that these particles are formed because of the different carbonization behaviors of GO and lignin. Lignin has more hydrogen, sulfur, and oxygen than GO, degrades faster and has greater mass loss, which might result in phase separation between carbonated lignin and RGO, resulting in the generation of such particles. The thickness of the carbide film remained at about 1 mm (Fig. 4.21c), indicating that no obvious thermal expansion occurred in the carbonization process at 873 K. HR-TEM images of the carbide film show a hybrid structure consisting of some graphite carbon from RGO and amorphous

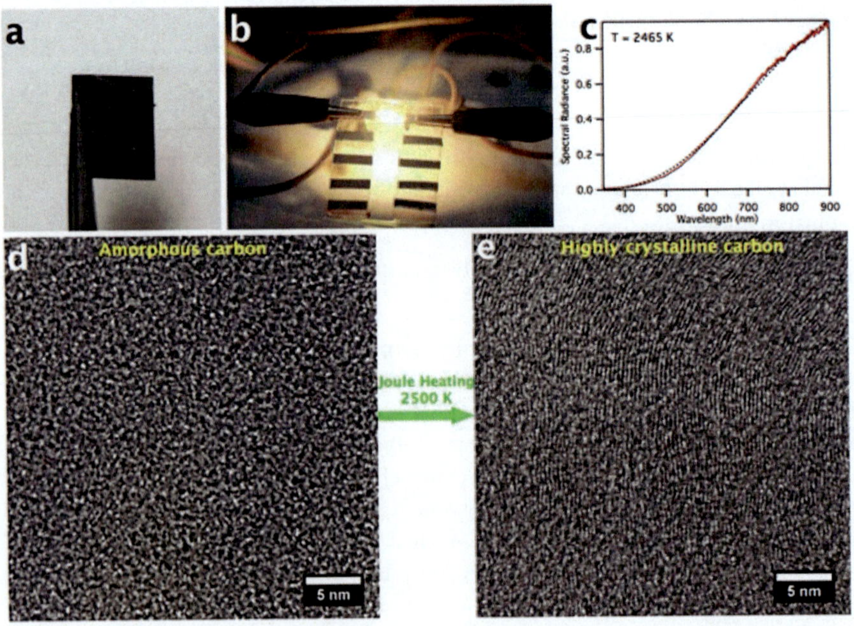

Fig. 4.20 **a** Carbon film of RGO lignin carbonized at 873 K; **b** photographs of RGO lignin carbon film strips in the process of high temperature joule heating; **c** spectral radiation measured in the process of joule heating; dashed lines represent fitted gray body radiation equations for temperature measurements. HR-TEM images of RGO lignin carbon films before **d** and after **e** joule heating

carbon from lignin (Fig. 4.21d). The graphite and amorphous domains were evenly mixed at tens of nanometers, manifesting that RGO and lignin were evenly mixed. The rapid Fourier transform of the region contained in the red box in Fig. 4.21a shows the combination of the amorphous halo (disordered) from the amorphous lignin carbon and the spots from the ordered, crystalline RGO (Fig. 4.21e), while merely the amorphous halo can be detected in the green box region (Fig. 4.21f).

This paper describes a new and simple high-temperature joule heating procedure for converting lignin into graphite carbon material with unprecedented high electrical conductivity. Joule heating depends on the internal resistance of the RGO lignin carbon film and is self-heating by introducing direct current. Under the direct current of 3.15A, the composite film could be heated to about 2500 K in 1 h to induce graphitization of lignin-based carbon. After graphitization, the lignin base graphite carbon exhibits ultra-high conductivity of 4500 S cm^{-1}, which is more than 700 times higher than the graphite carbon carbonized at 873 K. The results demonstrate the

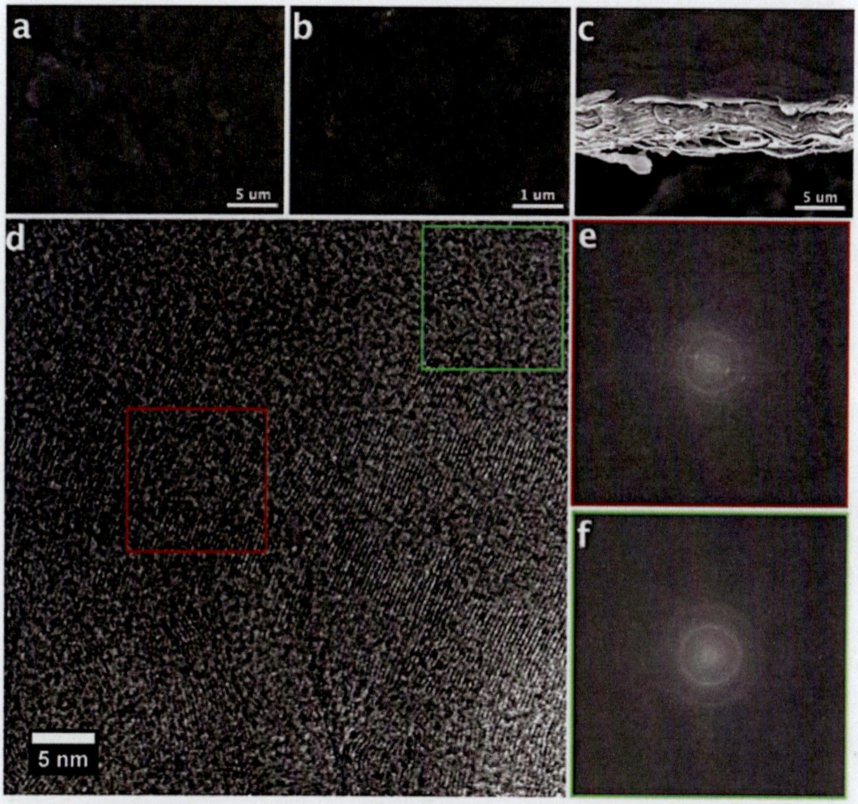

Fig. 4.21 Morphological characteristics of RGO lignin carbon film after high temperature joule heating **a**, **b** SEM image of surface; **c** SEM image of cross-sectional morphology; **d** HR-TEM image; **e**, **f** From the FFT pattern of the box shown in **d**

effectiveness of joule heating in converting bio-based materials into high-graphite carbon, and it is anticipated that the technology could be extended to other kinds of biomasses (e.g., chitin, cellulose, and even wood).

4.2.5 *Preparation of Carbon Composites by HTS*

(1) Rapid welding of nanocarbon coating on glass fiber by electrothermal shock method

With the fast development of nanomaterials fabrication, large-scale production of nanomaterials needs advanced technology of fabrication to combine nanomaterials with different materials (polymers, ceramics, and metals) to realize mixing and coupling properties in practical applications. The assembly of nanomaterials onto macroscopic materials is usually along with the loss of special nanoscale properties, primarily because of poor contact between carbon nanomaterials and macroscopic bulk materials. Shang et al. [35] proposed a new concept of cross-scale manufacturing to deal with different materials at different length scales and successfully proved an electrothermal shock method to deal with both nanomaterials (e.g., carbon nanotubes) and macroscales (such as glass fibers). The prepared material has good adhesion and excellent mechanical properties and is suitable for emerging applications. The superior property and potentially lower cost of electrothermal shock technology provide a successive, roll-to-roll, energy-efficient, ultra-fast heating solution for mass manufacturing.

CNT/glass fiber was picked as the model system to study the cross-scale electrothermal fabrication technology proposed in Fig. 4.22a. The electrothermal shock strategy developed in this work physically binds carbon nanostructures by using ultra-fast joule heating-assisted welding. The junctions between the nanotubes control resistance, causing local temperatures to rise, and the high thermal conductivity of the nanotubes rapidly transfers heat to adjacent regions. The heater caused by CNT junction resistance can be heated at an extremely high rate in one second (>1000 °C/s) generating high temperatures (above 1000 °C). The resulting CNT networks are coated on the glass fibers and used as nanoheaters to melt the glass in the contact area. The melted glass can physically anchor these CNT networks, generating a strong bond between the glass fibers and the CNT and improving the toughness of the glass fibers. Because nanowelding merely occurs on the surface for a brief time, the glass fiber body structure remains intact under ultra-fast electrothermal shock and retains the original mechanical performances.

This work shows that the carbon nanotube network can be tightly welded to the glass fiber while maintaining the typical properties of carbon nanotubes with good mechanical adhesion. Compared with the original glass fiber, the resulting CNT/glass fiber composite retains similar mechanical properties (stress and tensile modulus) and has higher toughness. The increase in toughness is primarily because of the strong anchoring CNT network which enhances the ductility of glass fibers. The

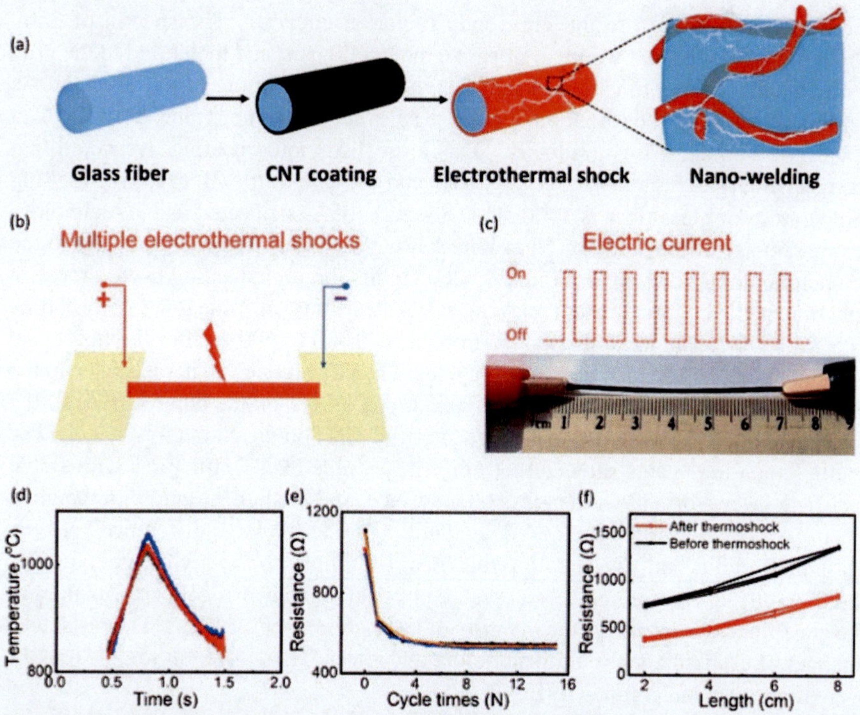

Fig. 4.22 Electrothermal shock for cross-scale fabrication of composite materials with multiple length scales **a** schematic diagram of design principle of electrothermal impactor; **b** schematic setup of the electrothermal shock procedure; **c** photographic image of a carbon nanotubule-coated fiberglass filament with two electrodes; **d** temperature–time relationship of carbon nanotubes/glass fibers during electrothermal shock in 1.5 s; **e** each electrothermal shock causes a change in resistance. The length of the coating CNT/glass fiber is 6 cm; **f** resistance of carbon nanotubes/glass fibers of different fiber lengths before and after electrothermal shock

design principles based on the CNT/fiberglass model can also be extended to other carbon nanomaterials and macrohost materials with melting temperatures.

Figure 4.22b shows a schematic diagram of the electrothermal shock nanowelding process. By utilizing pulse power, an electric current will flow through the conductive carbon layer to generate heat. By adjusting the current, the joule heat generated by the conductive carbon layer can increase the partial temperature and make the base material locally melt [36]. To avoid thermal degradation, materials like conductive carbon coatings and substrates must be protected from prolonged exposure by joule heating. Multiple electrothermal shocks can offer a fast high temperature environment to achieve nanowelding. The photograph exhibits a carbon nanotube/glass fiber (CNT/glass fiber) electrothermal shock device (Fig. 4.22c). A current with multiple cycle times is applied directly to the filament through a metal clip. Figure 4.22d exhibits the temperature–time relationship of the model based on CNT/fiberglass filament (6 cm long) at 0.105 A current. The temperature is calculated by fitting the emission spectrum to the gray body radiation equation. The spectra of 350–950 nm were collected by optical fiber (diameter 400 μm) and then fitted to the gray body radiation curve to calculate the joule heating temperature of the filament. As exhibited in Fig. 4.22d, the temperature reaches its maximum value at 1050 °C at 0.8 s and then decreases gradually. During the 1.5 s joule heating, we noted that the CNT/fiberglass filaments began to melt and become brittle. Because the melting temperature of glass fiber is 1100 °C, 0.8 s at 0.105A is selected as the technological parameter of the study of 6 cm long CNT/glass fiber filament according to the temperature and time curve. Figure 4.22e exhibits the joule heating cycle time as a function of CNT/glass filament resistance. Each current "on" time is 0.8 s, "off" time interval is 50 s. The large resistance drops from 1000 to 600 Ω after the first cycle and stabilizes at 530 Ω for the next 15 cycles. The large reduction in resistance in the first heating cycle may be owed to the oxidation of CNT on the outer surface of the CNT coating and partial thermal degradation of CNT in air. Stable resistance after multiple heating cycles showed that uniformly welded CNT structures with stable resistance were formed on the glass fiber surface, and the bulk structure in the glass fiber was not affected by thermal degradation. Due to the large difference between glass fiber radius and CNT coating thickness (<50 nm), so the high temperature of the fiber surface has minimal effect on the physical and mechanical performances of the glass fiber, thus ensuring the integrity of large glass fiber filaments. The resistance changes of CNT/fiberglass filament before and after HTS at different sample lengths were measured and compared (Fig. 4.22f).

The welded CNT/glass fiber has good mechanical bonding properties. As shown in Fig. 4.23a, CNTs can be injected into and fixed in glass fiber, and the welded carbon nanotubes can withstand intense ultrasonic action in aqueous solution without falling off. Figure 4.22b exhibits that after 3 h of cumulative ultrasonic treatment in water, there is no obvious change of color in the welded sample, however, the control sample without electric shock shows dark black in aqueous solution. The transparent solution showed no loss of the CNTs in water and showed a great bond between the glass fibers and the CNTs after intense ultrasonic treatment, while the black indicated a large number of carbon nanotubes falling into the water. After removal and drying, the welded CNT/glass fiber filaments remained black, but the control sample changed to white of the original fiberglass. Figure 4.23c–k shows the morphology of glass fibers before and after CNT welding. The surface morphology of raw glass fiber and CNT/glass fiber before and after electrothermal shock treatment was compared (Fig. 4.23c–e). The original glass fiber has a smooth surface and circular cross-sectional structure. After immersion in CNT ink and drying, the CNT nanonetwork is deposited on the glass fiber surface, forming a successive electron path to permit current to flow. After electrothermal shock, no pinhole was observed and CNT coating became dense. No morphological changes or structural damage were observed on the CNT/glass fibers after electrothermal shock, confirming that the welding procedure occurred primarily on the fiber surface. Use ultrasonic cleaning to remove unwelded CNT (Fig. 4.23f). By comparing the color changes of the two samples (Fig. 4.23b), it was concluded that the carbon nanotubes were welded into the glass fiber successfully and mechanically bonded to the glass fiber, with merely a small number of carbon nanotubes falling off the surface. The high-power SEM image (Fig. 4.23g, h) shows the structure of carbon nanotubes welded on the glass fiber after electrothermal shock. After high temperature impact, the cross section of the fiber bundle remains round and the diameter does not change (Fig. 4.23i–k) and maintains the overall structure. In cross sections, the welded CNTs can be identified and the cribbed profile of the glass fibers cannot be detected, further demonstrating that the glass fiber melting and carbon nanotube welding occur partially on the fiber surface and cause minimal thermal degradation and structural changes to the bulk glass fiber structure.

Figure 4.24a shows the TGA section for welded CNT/fiberglass and raw fiberglass. The TGA is heated at a constant flow speed of 10 °C min^{-1} over a temperature range of 20–1100 °C. The weight loss of raw glass fiber is primarily because of water removal and sizing decomposition. The weight loss of welding CNT/glass fiber is primarily because of CNT decomposition above 400 °C. In accordance with the weight comparison in Fig. 4.24a, welded CNT accounts for 3% of the overall CNT/glass fiber weight. Figure 4.24b exhibits the Raman spectra of welded and unwelded CNT/glass fibers. Compared with the unwelded CNT/glass fiber, the welded CNT/glass fiber retains great carbon characteristic peaks (D and G peaks), but the strength is lower. Raman spectroscopy results show that high temperature electrothermal shock has little effect on the properties of carbon nanotubes. We compared the welding CNT/glass fiber and raw glass fiber mechanical tensile properties (Fig. 4.24c–f). The results showed that after welding of CNT/glass fiber on the tensile strength, modulus, and toughness with the original glass fiber showed similar mechanical properties, this shows that the electric shock does not cause damage to the side effects of the original mechanical properties of CNTs. On the other hand, it is further confirmed that the welding between CNTs and glass fiber primarily occurs on the glass surface without affecting the overall properties. Therefore, the "similar" properties meet the goal of this work; that is, electrochemical shock can make CNT bond well to the glass fiber without compromising the overall mechanical properties. Figure 4.24c exhibits the stress–strain curves of two typical samples, which were tested at 25 °C at a tensile rate of 2 mm min^{-1} with a sample length of 5 cm. The tensile strength and modulus of glass fiber are 1.24 and 38.60 GPa, respectively. The tensile strength and modulus of welded CNT/glass fiber are 1.19 and 34.50 GPa, respectively. In particular, the welding toughness of CNT/glass fiber (2.61 GJ m^{-3}) was higher than that of glass fiber (2.29 GJ m^{-3}). The increased toughness can be owed to CNT welded on glass fibers, which form elastic nanonetworks and provide great structural integrity to improve mechanical robustness.

The above introduces a concept of cross-scale fabrication to deal with composite materials with multiple length scales and successfully demonstrates an electrothermal shock method to deal with nanomaterials (CNT) and macromaterials (glass fiber) with great adhesion while keeping great mechanical properties. Electrothermal shock utilizes high temperatures generated by joule heating of carbon materials. The CNT coating formed on the glass fiber surface acts as a nanoheater to melt the glass in the contact area at a heating speed of more than 1000 °C/s. The melted glass can physically anchor the CNT network, form a strong bond between the glass fiber and

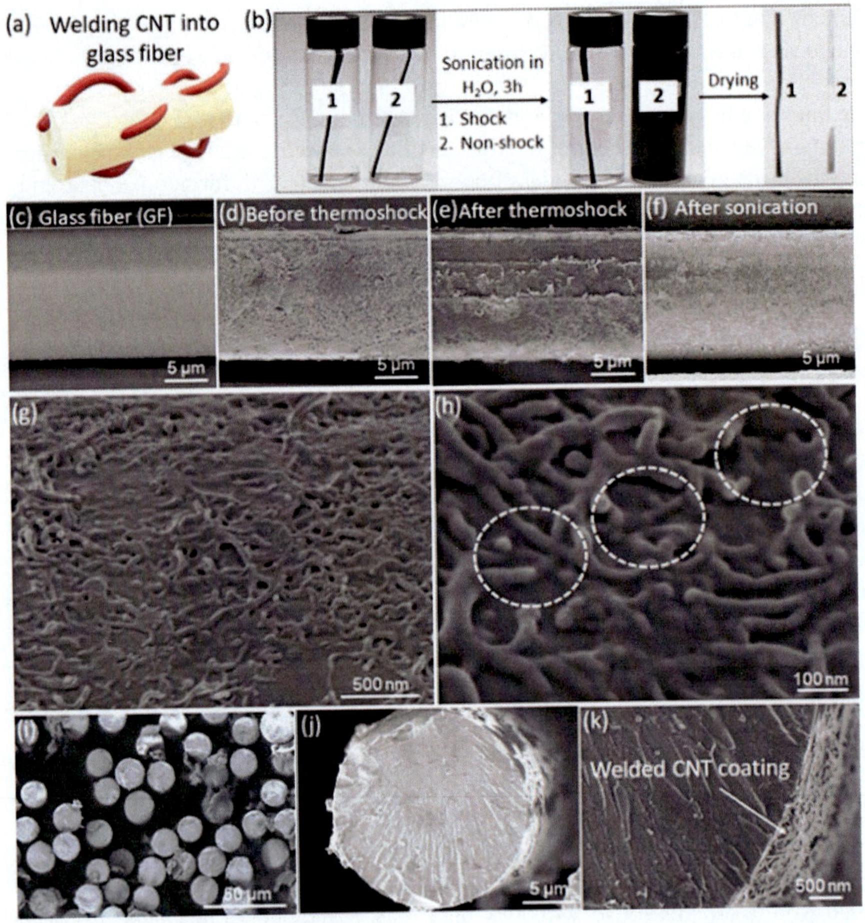

Fig. 4.23 Morphological features **a** schematic diagram of welded CNT/fiberglass structure; **b** photographs of non-impact and impact CNT/glass fiber before and after ultrasonic treatment in water (t = 3 h). SEM images of glass fiber surface welded CNT were amplified by **c** raw glass fiber; **d** unwelded CNT/glass fiber (before HTS); **e** welded CNT/glass fiber (after HTS); **f** welded CNT/glass fiber side view; **g–h** after ultrasonic cleaning to remove excess CNT; **i–k** cross-sectional diagram of welded CNT/glass fiber by CNTs forming nanonetworks and injecting into glass fiber surface

CNT, and improve the toughness of the glass fiber. After superfast electric shock, the structure of glass fiber remains intact and the original outstanding mechanical performance is maintained. The superior property and potentially lower cost of electrothermal impact technology provide an ultra-fast, successive, energy-efficient, low-cost roll-to-roll, and mechanically robust procedure that offers a heating solution with great promise for cross-scale fabrication of materials across multiple length scales.

(2) Strong CNT composite fibers: strong resistance to protonation, oxidation, and ultrasonic waves

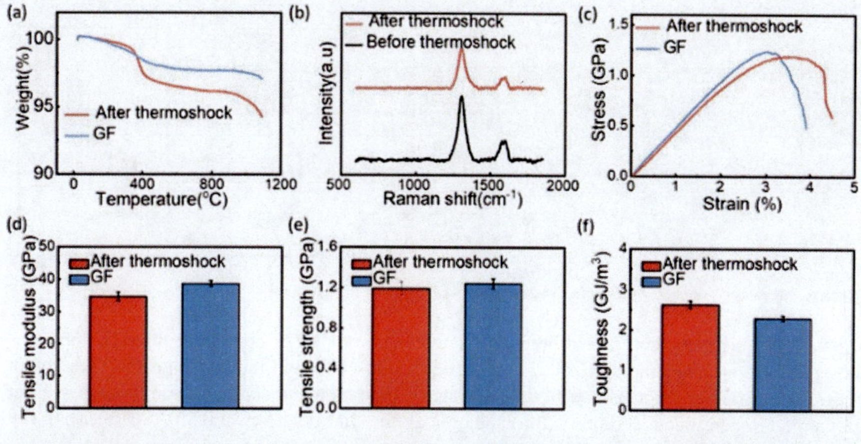

Fig. 4.24 Material and mechanical properties of CNT/glass fiber by high temperature electrothermal impact welding **a** TGA of raw GF and welded CNT/glass fiber; **b** Raman spectra of welded and unwelded CNTs/glass fibers; **c** stress–strain curves of raw glass fibers and welded CNTs/glass fibers; **d–f** Comparison of tensile modulus, tensile strength and toughness of raw glass fibers and welded CNTs/glass fibers

Carbon nanotube (CNT) fibers have excellent potential in high-performance fibers field. But the poor coupling between tube bundles results in poor structural and mechanical stability under strong acid, ultrasonic, and high-temperature oxidation conditions, which limits the practical application of CNTs in extreme environments. Li et al. [37] prepared strong CNT/C composite fibers with highly oriented and dense structure by ultra-fast joule heating and tensile annealing method. Because of the carbon-bonded structure, composite fibers exhibit better resistance to structural damage under strong acid, ultrasonic, and high temperature oxidation. Compared with the original CNT fiber, the fracture load, strength, and modulus of the composite fiber are increased by 320%, 354% (2.3 GPa), and 667% (60 GPa), respectively. In addition, the composite fiber has low density (1.48 g cm^{-3}) and great flexibility. These combination characteristics can widen the application of CNT/C composite fibers in lots of fields.

The composite is prepared as follows: CNT fibers collected on a winding machine are connected at one end to a rotating motor. CNT fibers are successively drawn through PAN/DMF solution at low rate. Meanwhile, the prepared CNT/PAN composite fibers were collected on a rotating spindle at low rate. At last, the produced composite fibers are placed in a vacuum furnace to remove residual solvents. (Fig. 4.25a).

Continuous preparation of CNT/C composite fibers is carried out in a custom-made quartz tube furnace that allows argon gas to be pumped through to prevent oxidation of CNT in the process of high temperature tensile annealing. Figure 4.25b shows the peroxide and carbonization settings of CNT/PAN composite fibers. The CNT/PAN composite fibers are fixed to one end of the quartz tube, passed through the carbonization zone, passed through the movable pulley, and then connected to a

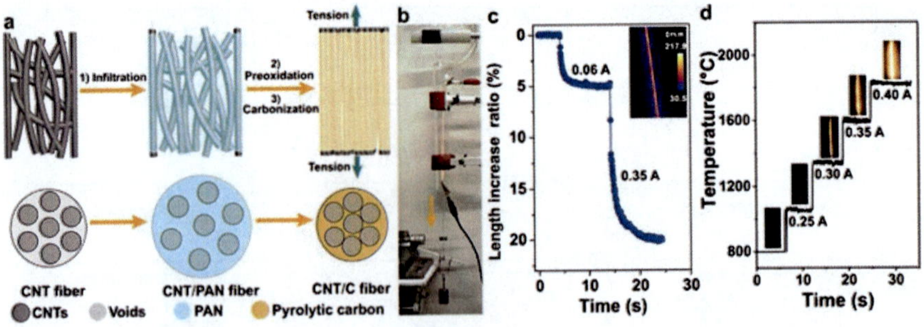

Fig. 4.25 **a** Schematic diagram of preparation procedure of CNT/C composite fibers; **b** apparatus for rapid tension annealing of CNT/C composite fibers; **c** by rapid tension annealing, the length of the composite yarn increases with time during the oxidation and carbonization processes. Inset are thermal images taken at 0.06 A (top) and optical fiber images taken at 0.35 A (bottom); **d** The temperature shall be measured with a colorimetric thermometer at an applied current of 0.25–0.40 A. The inserts showed that the fibers became brighter as the temperature and current increased

rotating motor for collection. A weight (100 g) suspended below the movable pulley is used to stretch and align the composite fibers, which is important for improving mechanical properties.

CNT/PAN composite fiber was pyrolyzed by two-step procedure. Firstly, CNT/PAN composite fibers were stabilized in air by applying a lower current (0.06 A) for different times (5 s, 10 s, 20 s, 30 s, 1 min, 2 min) under 1 N tension. After the power is cut off, the composite fibers are cooled to room temperature. For comparison, the thermal stability of CNT/PAN composite fibers was performed in an air oven at 220 °C for 2 h at 1 N. Secondly, in an atmosphere protected by argon, the applied current is increased until the fibers turn red or even incandescent. The temperature of CNTs increases rapidly, leading to the pyrolysis of polyacrylonitrile in CNT fibers. The influence of impressed current (0.25 A, 0.30 A, 0.35 A, 0.40 A) and carbonization time (2 s, 5 s, 10 s, 15 s) on carbonization was studied, and the conditions for the fiber to achieve the best properties were determined.

The morphology of raw CNT fiber, CNT/PAN, and CNT/C composite fiber was characterized. Compared with CNTs, a CNT fiber with 55 mm thick can be obtained by twisting the carbon nanotube ribbon (Fig. 4.26a, d). The cross-sectional SEM images show that there are a lot of voids in the original CNT fibers (Fig. 4.26g). After filtration, the diameter and twist angle of CNT/PAN fibers increased to 63 mm and 19.5°, respectively (Fig. 4.26b). In addition, few voids were found in CNT/PAN fibers because of PAN infiltration (Fig. 4.26e, h). The results exhibit that the diameter of CNT/C composite fiber is reduced to 50 mm and the deviation angle is as small as 8.6° (Fig. 4.26c). As a result of the applied tension, the arrangement of nanotubes in the CNT/C fiber is obviously enhanced (Fig. 4.26f). Holes can be observed in the cross section of the fiber, but the holes occupy a much smaller area than the original fiber (Fig. 4.26i).

The mechanical properties of the sample are tested, and the results are shown in Fig. 4.27. Due to loose structure, the strength of the original CNT fiber is low,

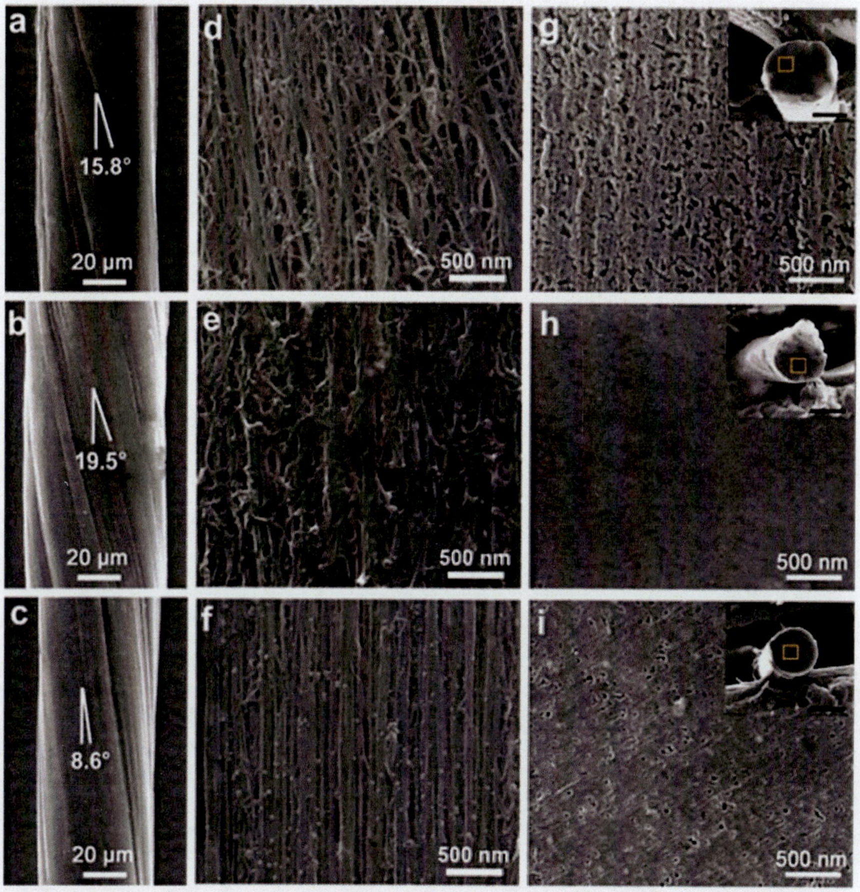

Fig. 4.26 **a–c** SEM images of raw CNT fibers, CNT/PAN, and CNT/C composite fibers at low magnification; **d–f** SEM images of raw CNT fibers, CNT/PAN, and CNT/C composite fibers at high magnification; **g–i** cross-sectional morphology of raw CNT fibers, CNT/PAN, and CNT/C composite fibers. The illustrations in **g–i** exhibit the cross-sectional morphology of raw CNT fibers, CNT/PAN fibers, and CNT/C fibers at low magnification, respectively. The scale is 40 μm

only 460 MPa. After impregnation with PAN/DMF solution, the tensile strength of CNT/PAN composite fiber was enhanced. When using 3 wt.% PAN/DMF solution, the tensile strength was slightly up to 840 MPa (Fig. 4.27a). With the decrease of annealing time (less than 10 s) and current (less than 0.35 A), the tensile strength of CNT/C fibers increases (Fig. 4.27b, c), which is because of the increased generation of amorphous carbon, which binds closely to the formed CNT bundles and inhibits the slip of the CNTs. Under the optimal joule heating time of 10 s and current of 0.35 A, the tensile strength and Young's modulus of the prepared CNT/C composite fiber are 2.2 GPa and 60 GPa, respectively. The fracture load is significantly increased due to the formation of larger beams and improved intertube connections. After peroxidation and carbonization of the fibers by percolation and subsequent rapid tensile annealing, the breaking load of the fibers increased to 4.2 N, an increase of 320% compared to the original CNT fibers (Fig. 4.27d).

The above describes a method for the rapid synthesis of CNT/C composite fibers with high strength, high hardness, and good performance even under extreme environmental conditions (e.g., high temperature oxidation, strong acid, and ultrasonic. The enhanced performances are primarily due to the great arrangement of nanotubes by tension annealing and the strong carbon bond between adjacent nanotubes. This method has high processing efficiency and can be used to continuously prepare CNT/C composite fibers and other CNT-based composite fibers. It can be used in high temperature applications, energy conversion textiles, artificial muscles, and other fields.

(3) Directionally electrothermally induced carbon deposition for welding carbon nanotube fibers

Zou et al. [38] proposed a simple method for welding CNT fibers by depositing carbon nanostructures at fiber connections. Electrothermal-induced deposition (ETID)

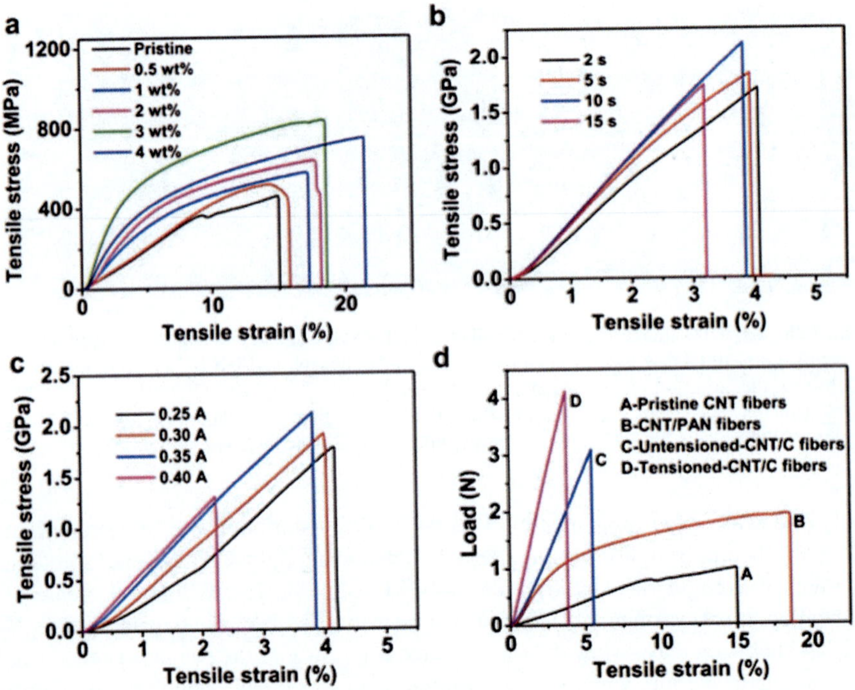

Fig. 4.27 **a** Typical stress–strain curves of original CNT fibers and CNT/PAN composite fibers under different impregnation solution concentrations; **b** stress–strain curves of CNT/C composite fibers under different carbonization times; **c** stress–strain curves of CNT/C composite fibers prepared under different carbonization currents; **d** stress–strain curves of raw CNT fibers, CNT/PAN composite fibers, unstretched CNT/C fibers, and stretched CNT/C fibers

procedure promotes thermochemical vapor deposition through current-induced joule heating. A variety of carbon structures are formed between overlapping fibers, and the nanofibers are covered by amorphous carbon and carbon nanowalls, thus achieving effective welding. The welded connection between the cross-fibers is much stronger than the original CNT fibers: The separation force of the measured connection is up to 460 mN, while the fracture force of the fibers is about 74 mN. After ETID treatment, the contact resistance is from > 120 down to 4.8 Ω. This robust welding can be used to join CNT fibers in parallel and cross configurations to form welded one-dimensional lines, two-dimensional networks, and three-dimensional frame structures.

ETID is an effective welding method for connecting two carbon nanotube fibers. Figure 4.28a schematically shows the method of welding two fibers into a straight line. The current is introduced by copper terminal clamp with 1–2 mm overlap. Different from traditional chemical vapor deposition, the huge contact resistance between fibers makes carbon deposition targeted. It occurs first at the junction and then around. After ETID treatment of up to 500 mA (within 500 s), a fibrous structure is formed around the fragment between the two copper clips (Fig. 4.28b). Due to the deposition produced by ETID, the total mass of both fibers is greatly increased. For instance, the mass was increased from 97 to 303 mg. Therefore, due to the deposition of a large number of carbon nanostructures between the fibers, it is difficult to distinguish the overlaps by naked eyes, as shown in Fig. 4.28b.

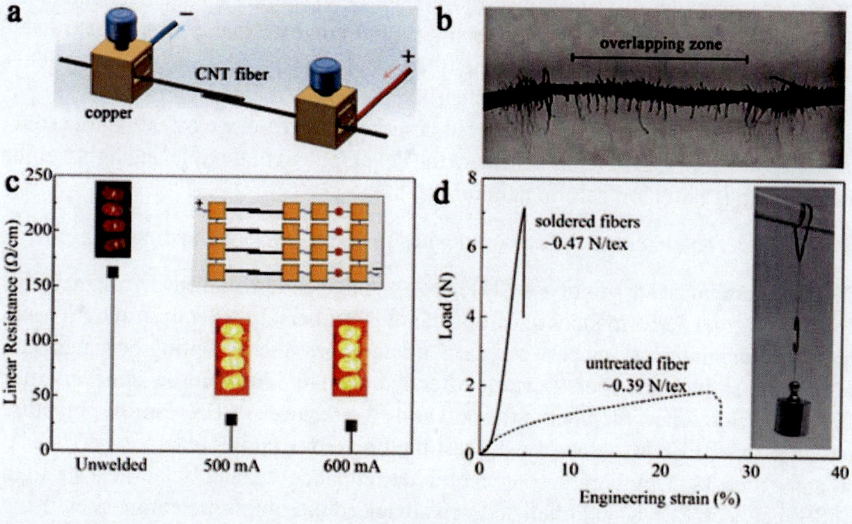

Fig. 4.28 Welding of two overlapping CNT fibers **a**, **b** wire welding diagram and a welding example; **c** resistance of untreated 500 and 600 mA welded optical fibers. The illustration compares the brightness difference of four LED lights connected to the optical fiber and shows the corresponding parallel circuit; **d** Tensile performances of untreated fibers and 500 mA welded fibers. The illustration exhibits that the welded ring of the carbon nanotube fiber can suspend a weight of 10 g

In order to exhibit the effect of ETID welding, four pairs of fiber contacts are arranged in the parallel circuit, and each pair of fiber contacts is connected with an LED light, as illustrated in Fig. 4.28c. The average linear resistance of the four pairs of untreated fiber contacts is 162.2 Ω/cm (the circuit is disconnected when measuring each pair of resistance), and the LED lamp is very dim by applying a voltage of 2.5 V across the entire circuit and connecting circuit. The average linear resistance of each pair of electrodes was significantly reduced to 28.9 Ω/cm after ETID treatment. This implies that the contact resistance of each pair results in a drop of 133.3 Ω/cm. When the entire circuit is turned on at 2.5 V, the LED lights become very bright. The situation was almost the same at 600 mA, with an average resistance of 24.5 Ω/cm.

ETID welding not only reduces contact resistance, but also provides a strong connection between the CNT fibers. Figure 4.28d compares the tensile tests of untreated fibers and fibers treated at 500 mA. It takes only 1.88 N to break untreated fibers, while welded fibers can withstand forces up to 7.15 N. To show reinforcement more clearly, the ends of the narrow CNT strip (1.2 mm wide by 46 mm long, overall mass 224 mg) were welded together to form a ring that could easily suspend a 10 g (up to 15 g) dangling object, as illustrated in Fig. 4.28d.

An electrothermal-induced deposition process is described above, which can be used as an effective method for CNT fiber welding. Targeted deposited carbon structures, containing carbon nanofibers and nanowalls, grow between CNT fibers and connect the fibers by forming welded joints with outstanding mechanical and electrical performances. A variety of sizes and programmable circuits were constructed from carbon nanotube fibers, proving the simplicity of ETID technology. This approach can be utilized to all kinds of carbon nanotube fibers and provides a simple method for constructing all carbon-based circuits. Therefore, we offer a viable solution to the key problem that has restricted the development of flexible and lightweight smart facilities based on carbon nanotubes.

(4) Resin/carbon nanotube fiber composites prepared by joule heating

The arrangement and assembly of CNTs are crucial to obtain high electrical, mechanical, and thermal performances of carbon nanotubes fibers. In order to surmount weak van der Waals interactions between carbon nanotubes, thermosetting polymers can be introduced into the fibers because effective strengthening can be achieved after thermal curing. A new member of the polyaryl ether ketone (PAEK) family, polyether ketone ketone (PEKK) shows the highest melting temperature, reaching 380 °C. At the same time, high temperatures are required to enhance the interfacial shear strength (IFSS) between PEKK and fibers to form advanced thermoplastic composites. Yang et al. [39] designed a more effective strategy to enhance IFSS due to the strong joule heat in CNT fibers. Under the electrical treatment, the surrounding PEKK begins to melt on the surface of the fibers and even seeps into the carbon nanotube assemblies. The results show that the interaction between interfaces is significantly enhanced. The maximum IFSS of the electrotreatment interface between CNT fiber and PEKK is 80.4 MPa.

Qiu et al. [40] used the electrocuring method to cure bismaleimide (BMI) impregnated CNT fibers to enhance intertube heat transfer. The aromatic BMI resin solidifies into a directional structure when an electric current is carried along the surface of the nanotubes and the strongest joule heat is sensed there. At the optimal curing current with strong joule heating and no BMI degradation, the intrinsic thermal conductivity of the fibers increased from 30 to 177 W m^{-1} K^{-1}, and the heat conductivity reached 374 W m^{-1} K^{-1} (sample length was 12 mm).

4.3 Application of HTS in Structural Regulation of Carbon Materials

HTS technology can not only be used to prepare carbon materials, but also be widely used to prepare and adjust the internal structure of materials to improve the performances of materials. The application of this technology in material structure control will be introduced below.

(1) MSEC tensile annealing reinforced CNT fibers

CNT fibers with high mechanical strength have become the focus of research in the industry of fiber. Nevertheless, weak, or even lack of connections between adjacent CNTs will cause a lot of intertube slippage when fibers fail, resulting in low mechanical strength. In addition, achieving large-scale rapid cross-linking between adjacent CNTs to prevent slip failures remains a great challenge. Song et al. [41] reported an ultra-fast continuous tensile annealing procedure to realize significantly enhanced tube alignment and strong covalent cross-linking of adjacent CNTs in milliseconds, thus greatly improving fiber properties. Under the applied tension, the CNT fibers were heated to high temperature (about 2450 °C) by joule heating and then annealed for 12 ms. Because of the fast electromechanical response of the fibers, an instantaneous nanotube rearrangement occurs when energized by forming cross-links that firmly connect adjacent CNTs, which can be attributed to a significant 2.9-fold (up to 3.2 GPa) and 4.8-fold (up to 123 GPa) increase in strength and modulus, respectively. The specific strength (2.2 N/tex) of the synthetic fiber is similar to that of PAN-based carbon fiber, and the specific conductivity is higher than that of PAN-based carbon fiber. The resulting strong cross-linking and high-density structure confer obviously enhanced thermal stability of the fibers under high temperature oxidation atmosphere. In addition, a successive tensile annealing procedure was designed to mass produce high-performance fibers with an average strength of 2.2 GPa. The high toughness, lightweight, continuity, and excellent mechanical and electrical performances of carbon nanotube fiber will promote its large-scale application.

Figure 4.29a diagonally exhibits the welding procedure of CNT fibers caused by ultrafine joule heating under tension. A high voltage pulse is utilized to the raw fiber tensioned by a weight. Because of the bias angle on the carbon nanotube fibers, the stress applied by weight produces transverse stress to bind the carbon nanotubes and CNT bundles tightly together, which helps align the CNTs and form cross-links between the CNT and CNT bundles. The improved arrangement and cross-linking in the larger CNT bundles can prevent the slip of CNTs and effectively enhance the strength of CNT fibers. Temperature and heating speed are controlled by adjusting the input voltage source at millisecond resolution. CNT fibers have fast electrothermal response because of their ultra-small heat capacity per unit area, which can be heated to incandescence in milliseconds under vacuum and cooled in milliseconds. In an argon-filled quartz tube, the growing CNT fibers tensioned by hanging weight were electrically heated to more than 2000 °C within 12 ms by a high-pressure pulse of 70 V (Fig. 4.29b). When energized, the fiber immediately extends over 10% without hysteresis (Fig. 4.29c), and the diameter decreases rapidly, corresponding to the rapid increase of dark string span of the diffraction fringe formed (Fig. 4.29c illustration). The results showed that the primary CNTs fibers were densified by joule heating and tensile annealing to form a tighter arrangement structure.

The mechanical properties of the samples were tested. The original CNT fibers exhibited relatively low breaking load (0.16 N), strength (1.1 GPa), and specific strength (0.72 N/tex) because of their loose stacked structure and lack of effective bonding between adjacent CNTs to restrict the fibers from sliding in mechanical tests. Nevertheless, the strong interlaminar structure formed after joule heating and tensile annealing can effectively enhance the interaction between tubes, thus significantly improving the bearing capacity of fibers. The CNT fibers annealed at higher

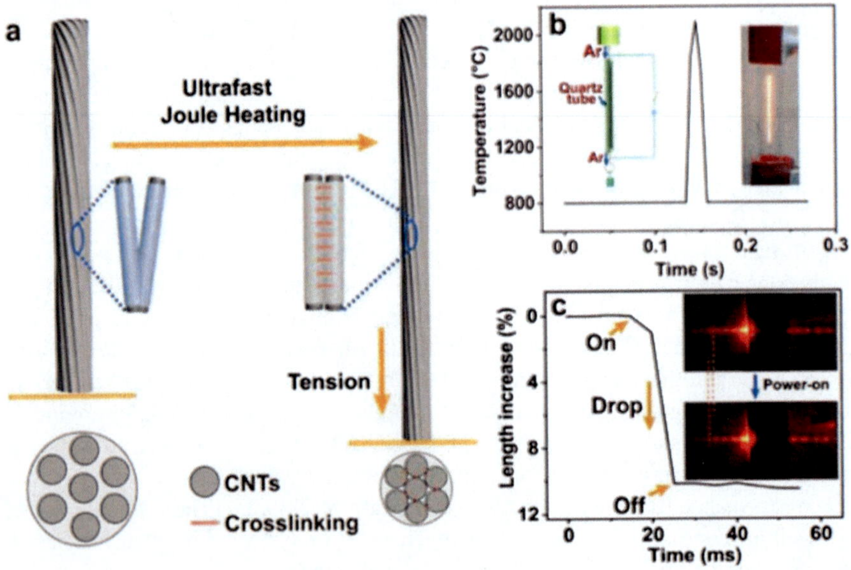

Fig. 4.29 **a** Schematic diagram of the welding procedure of CNT fibers caused by ultrafine joule heating under tension; **b** temperature at the midpoint of the CNT fiber, heated by a 70 V voltage pulse for 12 ms. The inset shows the joule heating setup and lighting images during joule heating above 2000 °C; **c** the length of CNT fibers was increased by applying 70 V voltage pulse for 12 ms and 0.05 N annealing tension

applied tension exhibited higher mechanical properties (Fig. 4.30a) due to their better nanotube arrangement and stronger cross-links between adjacent CNTs within the larger CNT bundle formed. At the applied pulse voltage of 70 V, the breaking load and specific strength of the fibers annealed within 12 ms were obviously increased to 0.34 N and 2.2 N/tex, which were 117% and 206% higher than those of the raw CNT fibers (Fig. 4.30b, c). The tensile strength and modulus of annealed fibers increased 2.9 times (up to 3.2 GPa) and 4.8 times (up to 123 GPa), respectively. In addition, similar results can be obtained in load–displacement curves recorded utilizing nanoindenter (Fig. 4.30d).

Based on the fast electromechanical response characteristics of CNT fibers, an ultra-fast joule heating and tensile annealing method was proposed to realize covalent welding of adjacent carbon nanotubes in the fibers within milliseconds. After annealing at high temperature (2450 °C) at only 12 ms under tension, the CNT in the fibers can be highly aligned and tightly packed together to form thicker bundles through C–C cross-linking, thus obviously enhancing the mechanical properties and electrical conductivity of the CNT fibers. The specific strength of the synthetic fiber is increased to 2.2 N/tex (249%), which is comparable to pan-based carbon fibers.

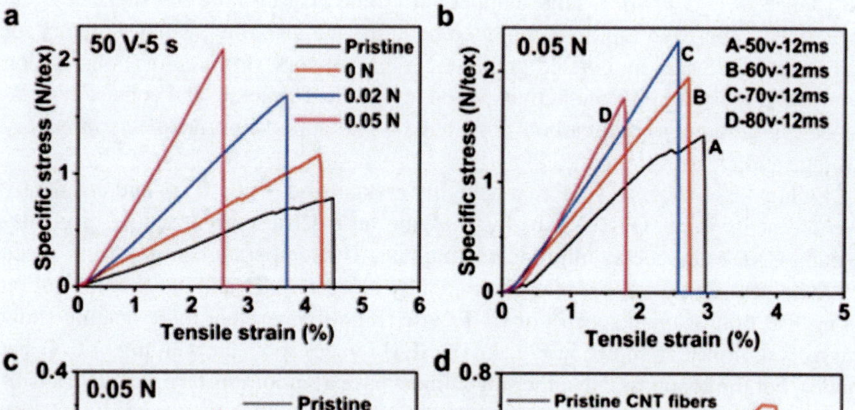

Fig. 4.30 **a** Stress–strain curves of raw and annealed CNT fibers under different tensile tension; **b** specific stress–strain curves of fibers annealed at 0.05 N at different voltages for 12 ms; **c** load–strain curves of primitive fibers and fibers annealed for 12 ms at a pulse voltage of 70 V at 0.05 N tension; **d** load–displacement curves of raw CNT fibers and fibers annealed for 12 ms at 70 V pulse voltage

The specific conductivity of the annealed fibers reached 1060 S cm^2 g^{-1}, which was higher than that of PAN-based fibers. In addition, the synthesized fibers also exhibit high toughness and significantly enhanced thermal stability under high temperature oxidation atmosphere. Therefore, the prepared annealed carbon nanotube fibers can be used as a great solution for high strength, lightweight, and conductive materials.

(2) Ultra-fast joule heating carbon welding

Carbon nanomaterials show excellent electrical and mechanical performances, but these excellent performances are often destroyed because nanomaterials are assembled into bulk structures. This scale problem restrains the utilization of carbon nanostructures and can be owed to poor physical contact between nanostructures. In order to address this challenge, Yao et al. [42] proposed a new technique to construct a three-dimensional interconnected carbon matrix by forming covalent bonds between carbon nanostructures. Carbon nanofiber (CNF) films were heated to a temperature greater than 2500 K at a heating speed of 200 K min^{-1} by high-temperature joule heating. The adjacent CNFs and graphite carbon bonds were fused together to form a three-dimensional successive carbon network. The volume conductivity of carbon matrix increases by four orders of magnitude to 380 S cm^{-1}, and the sheet resistance is 1.75 Ω sq^{-1}. High-temperature joule heating not only makes carbon materials graphitized rapidly at high temperature, but also offers a new strategy for constructing covalently bonded graphite carbon networks from amorphous carbon sources. Due to its high conductivity, good mechanical structure and corrosion resistance, 3D interconnected carbon films have broad application prospects in energy storage and electrocatalysis.

Figure 4.31a shows a CNF that is highly crystallized (Fig. 4.31a) and covalently welded at its joints (Fig. 4.31b) by applying an electric current to the originally wound CNF to induce an ultrafine heating rate. The temperature and heating speed of joule heating technology are controlled by electric field, and its time resolution is millisecond. Raman spectra of CNFs after high-temperature joule heating show high crystallinity, as shown in Fig. 4.31b. SEM images show the welding of CNFs at nodes. For the original CNFs, the single fibers have a smooth surface and are translucent because of their amorphous structure. In the original CNF matrix, overlapping nanofibers have physical contact with adjacent nanofibers, but the van der Waals interaction is weak. After joule heating, smooth fibers become rough and physical contact is welded at the nodes to form a complete optical fiber network (Fig. 4.31d). By joule heating, the amorphous wound CNFs become highly graphitized CNFs, which are welded at the joints. The highly graphitized covalent interbond structure effectively reduces the contact resistance and facilitates rapid electron transfer. In addition, the mechanical properties of the welded CNFs are improved because the covalent welding is carried out at the joint of the fibers, which makes the fibers not easy to slip.

After joule heating, the CNF film formed a highly crystalline carbon structure longitudinally, as shown by the arrow in Fig. 4.32a, with graphite carbon layers longitudinally. During high temperature joule heating, hydrogen and nitrogen are removed from the PAN-derived carbon structure, which facilitates cross-linking into

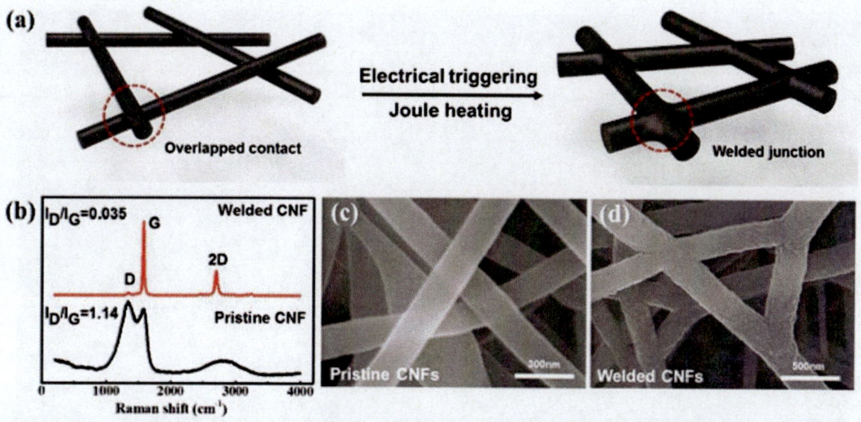

Fig. 4.31 Welding process schematic diagram of CNF and selected characterization results **a** schematic diagram of CNFs carbon welding procedure triggered by joule heating; **b** Raman spectra of original CNF and joule-heated CNF; **c, d** SEM images of raw CNF and welded CNF

a crystalline graphite structure. Figure 4.32b shows the TEM image of the joule-heated carbon nanofiber welded joint. Highly crystalline carbon layers can be clearly observed at the junction points connecting adjacent CNFs. In order to understand the welding behavior under joule heating treatment, we controlled the temperature caused by joule heating and observed the morphological evolution of CNFs, as exhibited in Fig. 4.32c–f. Due to its amorphous structure, the original CNF film exhibits smooth, wound fibers with a diameter of about 200 nm and translucent optical properties. After the joule is heated to 1500 K, the fibers begin to stick together in parallel, thus forming a "Y" shape, and cross, forming a "+" shape. At 1500 K, there are still clear boundaries between these fibers, indicating that the welding between carbon nanofibers is incomplete. At 1800 K, the fiber surface becomes rough and the nanofibers begin to merge and weld together, resulting in blurred boundaries between the fibers. At 2300 K, the knots are fully welded together with no distinct boundaries. It is worth noting that high temperature joule heating can weld not only adjacent cross fibers but also parallel fibers, thus making a wider range of fiber diameters. The microstructure of CNFs at different temperatures during joule heating confirmed the melting of CNFs between 1500 and 2300 K. In this range of temperature, the amorphous CNFs undergo carbonization and graphitization processes, which eliminate the defective atoms and form C–C bonds at the nanofiber junctions. The nanofibers were further annealed at high temperature above 2400 K and fused into a highly crystalline graphite carbon matrix.

The high graphitization and joint welding of the CNF films are beneficial to the electron transport in the bulk structure and significantly enhance the conductivity of the carbon fiber films. The conductivity of CNF film increases 30,000 times from 13.3 mS cm^{-1} to 381 S cm^{-1}, and the film resistance is 1.75 Ωsq^{-1} (Fig. 4.33a). This calculation was made without subtracting the porosity of CNF films with a density of about 0.2–0.3gcm^{-3}, confirming the superior performance

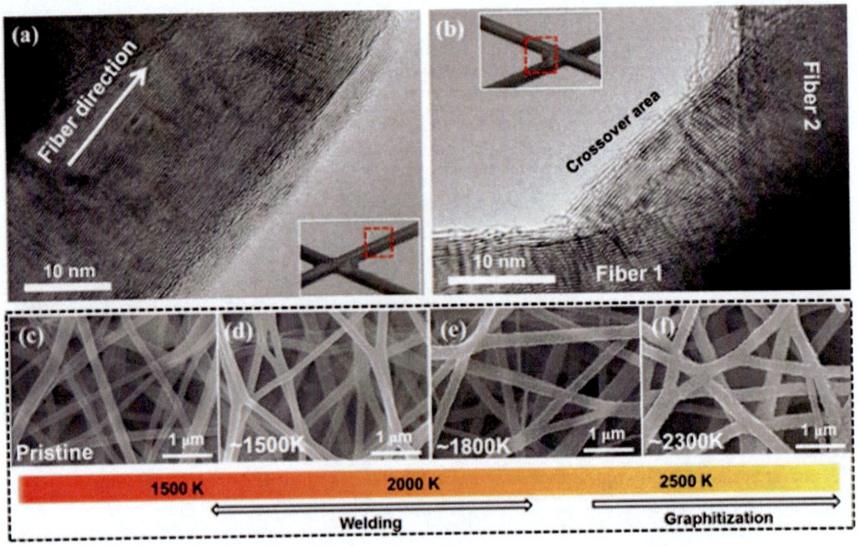

Fig. 4.32 Morphology characterization of welded CNF structure **a** HR-TEM images of joule-heated CNF; **b** HR-TEM images of the joints between the joule-heated CNFs; **c–f** SEM images show the morphological evolution of the original CNF and CNF joule heated at 1500 K, 1800 K, and 2300 K

of blocky, three-dimensional interconnected carbon substrates with welded joints. In order to understand the character of node welding in improving electrical performance, the conductivity of individual fibers was measured using a shadow mask technique (Fig. 4.33b). A single optical fiber is ultrasonic from the CNF film and transferred to a silicon wafer. The 100 nm gold electrodes are deposited on both ends of the fiber through a shadow mask (Fig. 4.33b inset). The conductivity of low-temperature carbonized single fiber CNF (s-CNF) is 4.67 s cm^{-1}. After high joule heating, the conductivity of single fiber (j-CNF) increased by about 180 times by dual probe measurement, reaching 822 S cm^{-1}. Figure 4.33c shows the conductivity of the original CNF film, welded CNF film, single CNF fiber, and single CNF fiber after joule heating. After joule heating, the electrical conductivity of both films and single fibers increased significantly, but the amplitude was different. The conductivity of CNF film changes by a factor of 30,000, while the conductivity of a single nanofiber increases merely by a factor of 180. We speculate that welding nanofiber junctions in CNF films is also crucial for improving the conductivity of welded CNF films.

The high temperature joule heating approach is introduced above to realize covalent welding of adjacent fibers in 3D CNF matrix to solve the scaling problem of carbon nanomaterials. CNF film is heated to high temperature at superfast speed (>2500 K) (200 K min^{-1}), thus significantly improving the overall conductivity. The bulk conductivity of the carbon structure increases by four orders of magnitude to 380 S cm^{-1}, and the chip resistance is 1.75 Ω sq^{-1}. This work shows that the high temperature joule heating can rapidly graphitize carbon materials at high

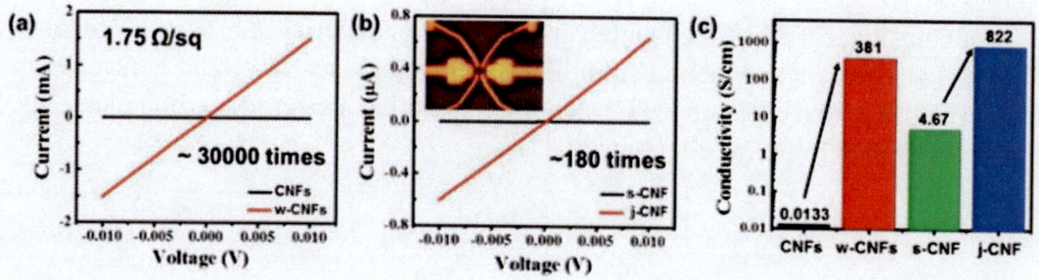

Fig. 4.33 Electrical conductivity of a welded CNF (w-CNFS) structure **a** conductivity measurement results of CNF film before and after high-temperature joule heating; **b** conductivity measurement results of single fiber (s-CNF) before and after joule heating (j-CNF); **c** conductivity of raw CNF film (CNF), welded CNF film (w-CNF), single fiber CNF (s-CNF), and joule-heated single fiber CNF (j-CNF)

temperatures and offers a new strategy for constructing covalent interconnect carbon networks from amorphous carbon. The researchers speculate that the method could be extended to other carbon materials (e.g., carbon nanotubes and graphene) by appropriate surface modifications, such as utilizing polymers to generate graphite carbon coatings on graphene and carbon nanotube networks, and forming covalently interconnected carbon structures through high-temperature joule heating methods.

Joule heating can also be used to weld graphene. Prof. Liu et al. [43] welded single macrographene components using electric joule heating. Graphene-based materials of different structures, containing graphene fibers, films and foams, are covalently welded together for high joint strength. This unique welding behavior is attributed to defects between graphene layers at the contact interface that promotes cross-linking. The simple joule welding strategy offers a new and reliable bonding approach for carbon materials, allowing macroscopic graphene materials to be directly assembled into complex three-dimensional configurations.

4.4 Summary

The application of HTS in carbon materials is introduced from two aspects above. HTS can be increased to ultra high temperature in a very short time, with low cost, high efficiency, green environmental protection, and other advantages. Compared with the traditional preparation methods, the carbon materials prepared by HTS technology are less prone to agglomeration, oxidation, and other phenomena, which can ensure the maximum performance of the material.

HTS is not only can be used in the preparation of graphene films, graphene fiber, graphene powder such as different forms of carbon materials, but also by the carbon material and the technology of inorganic non-metallic materials (e.g., glass fiber), composite, polymer materials (resin), for example, that takes both advantages of composite materials. HTS is used to adjust the internal structure of the material to form a new bond or configuration in order to obtain the best performance.

In conclusion, HTS has been widely used in carbon materials due to its advantages of fast, low cost, environmental friendliness, universal, scalable, and controllable. We can continue to try to apply HTS technology to other aspects of carbon materials, in order to obtain more excellent results.

References

1. Novoselov KS, Geim AK, Morozov SV et al (2004) Electric field effect in atomically thin carbon films. Science 306(5696):666–669
2. Zhang Y, Zhang L, Zhou C (2013) Review of chemical vapor deposition of graphene and related applications. Acc Chem Res 46(10):2329–2339
3. Amatucci GG, Badway F, Du Pasquier A et al (2001) An asymmetric hybrid nonaqueous energy storage cell. J Electrochem Soc 148(8):A930–A939
4. Krishnan D, Kim F, Luo J et al (2012) Energetic graphene oxide: challenges and opportunities. Nano Today 7(2):137–152
5. Eda G, Fanchini G, Chhowalla M (2008) Large-area ultrathin films of reduced graphene oxide as a transparent and flexible electronic material. Nat Nanotechnol 3(5):270–274
6. Wang Y, Chen Y, Lacey SD et al (2018) Reduced graphene oxide film with record-high conductivity and mobility. Mater Today 21(2):186–192
7. Marcano DC, Kosynkin DV, Berlin JM et al (2010) Improved synthesis of graphene oxide. ACS Nano 4(8):4806–4814
8. Chen H, Müller MB, Gilmore KJ et al (2008) Mechanically strong, electrically conductive, and biocompatible graphene paper. Adv Mater 20(18):3557–3561
9. Tu NDK, Choi J, Park CR et al (2015) Remarkable conversion between n- and p-type reduced graphene oxide on varying the thermal annealing temperature]. Chem Chemistry 27(21):7362–7369
10. Li T, Pickel AD, Yao Y et al (2018) Thermoelectric properties and performance of flexible reduced graphene oxide films up to 3,000 K. Nat Energy 3:148–156
11. Xiao N, Dong XC, Song L et al (2011) Enhanced thermopower of graphene films with oxygen plasma treatment. ACS Nano 5(4):2749–2755
12. Bong Lee H, Hyun Noh S, Hee HT (2021) Highly electroconductive lightweight graphene fibers with high current-carrying capacity fabricated via sequential continuous electrothermal annealing. Chem Eng J 414:128803
13. Hyun NS, Hun P, Wonsik E et al (2020) Graphene foam cantilever produced via simultaneous foaming and doping effect of an organic coagulant. ACS Appl Mat Inter 12(9):10763–10771
14. Hyun NS, Wonsik E, Jun LW et al (2019) joule heating-induced sp2-restoration in graphene fibers. Carbon 142:230–237
15. Park H, Lee KH, Kim YB et al (2018) Dynamic assembly of liquid crystalline graphene oxide gel fibers for ion transport. Sci Adv 4(11):2104
16. Pei S, Zhao J, Du J et al (2010) Direct reduction of graphene oxide films into highly conductive and flexible graphene films by hydrohalic acids. Carbon 48(15):4466–4474
17. Schniepp HC, Li JL, McAllister MC et al (2006) Functionalized single graphene sheets derived from splitting graphite oxide. J Phys Chem B 110(17):8535–8539
18. Mao W, Jin X, Wang H et al (2019) The association between resting heart rate and urinary albumin/creatinine ratio in middle-aged and elderly Chinese population: a cross-sectional study. J Diabetes Res 2019:1–7
19. Cheng Y, Cui G, Liu C, et al (2021) Electric current aligning component units during graphene fiber joule heating. Adv Funct Mat, 2103493
20. Stanford MG, Bets KV, Luong DX et al (2020) Flash graphene morphologies. ACS Nano 14:13691–13699

21. Luong DX, Bets KV, Algozeeb WA et al (2020) Gram-scale bottom-up flash graphene synthesis. Nature 577(7792):647–651
22. Wyss KM, Wang Z, Alemany LB et al (2021) Bulk production of any ratio (12)C:(13)C Turbostratic flash graphene and its unusual spectroscopic characteristics. ACS Nano 15(6):10542–10552
23. Algozeeb WA, Savas PE, Luong DX et al (2020) Flash graphene from plastic waste. ACS Nano 14(11):15595–15604
24. Wyss KM, Beckham JL, Chen W et al (2021) Converting plastic waste pyrolysis ash into flash graphene. Carbon 174:430–438
25. Advincula PA, Luong DX, Chen W et al (2021) Flash graphene from rubber waste. Carbon 178:649–656
26. Chen W, Li JT, Wang Z et al (2021) Ultrafast and controllable phase evolution by flash joule heating. ACS Nano 15(7):11158–11167
27. Krystek M, Pakulski D, Patroniak V et al (2019) High-performance graphene-based cementitious composites. Adv Sci 6(9):1801195
28. Toh SY, Loh KS, Kamarudin SK et al (2014) Graphene production via electrochemical reduction of graphene oxide: Synthesis and characterisation. Chem Eng J 251:422–434
29. Chen J, Yao B, Li C et al (2013) An improved Hummers method for eco-friendly synthesis of graphene oxide. Carbon 64:225–229
30. Kong W, Kum H, Bae S-H et al (2019) Path towards graphene commercialization from lab to market. Nat Nanotechnol 14(10):927–938
31. Cong C, Yu T, Saito R et al (2011) Second-order overtone and combination Raman modes of graphene layers in the range of 1690–2150 cm^{-1}. ACS Nano 5(3):1600–1605
32. Niilisk A, Kozlova J, Alles H et al (2016) Raman characterization of stacking in multi-layer graphene grown on Ni. Carbon 98:658–665
33. Rao R, Podila R, Tsuchikawa R et al (2011) Effects of layer stacking on the combination Raman modes in graphene. ACS Nano 5(3):1594–1599
34. Advincula PA, Luong DX, Chen W et al (2021) Ultrahigh-temperature conversion of biomass to highly conductive graphitic carbon. Carbon 178:649–656
35. Shang Y, Shi B, Doshi SM et al (2020) Rapid nanowelding of carbon coatings onto glass fibers by electrothermal shock. ACS Appl Mat Inter 12(33):37722–37731
36. Liu Y, Li P, Wang F et al (2019) Rapid roll-to-roll production of graphene films using intensive joule heating. Carbon 155:462–468
37. Li M, Song Y, Zhang C et al (2019) Robust carbon nanotube composite fibers Strong resistivities to protonation, oxidation, and ultrasonication. Carbon 147:627–635
38. Zou J, Zhang X, Xu C et al (2017) Soldering carbon nanotube fibers by targeted electrothermal-induced carbon deposition. Carbon 121:242–247
39. Yang X, Zhao J, Wu K et al (2019) Making a strong adhesion between polyetherketoneketone and carbon nanotube fiber through an electro strategy. Compos Sci Technol 177:81–87
40. Qiu L, Guo P, Yang X et al (2019) Electro curing of oriented bismaleimide between aligned carbon nanotubes for high mechanical and thermal performances. Carbon 145:650–657
41. Song YH, Di JT, Zhang C et al (2019) Millisecond tension-annealing for enhancing carbon nanotube fibers. Nanoscale 11(29):13909–13916
42. Yao Y, Fu KK, Zhu S et al (2016) Carbon welding by ultrafast joule heating. Nano Lett 16(11):7282–7289
43. Liu Y, Liang C, Wei A et al (2018) Solder-free electrical joule welding of macroscopic graphene assemblies. Mat Today Nano 3:1–8

Chapter 5
Application of HTS in Material Preparation and New Devices

Contents

5.1 Introduction

Traditional heating equipment mainly includes box-type resistance furnace, well resistance furnace, salt bath furnace, induction heating equipment, etc. The principle of box-type resistance furnace is that the electric current flows through and heats up the electric heating element arranged in the furnace and the parts are heated by convection and radiation. But the temperature difference in the furnace is large and the operation is not convenient, especially the large box-type resistance furnace, workers in the operation of the labor intensity is large. The working principle of the well resistance furnace is the same as that of the box-type resistance furnace. The furnace door is located on the top of the furnace, and the workpiece is vertically suspended in the furnace by special spanner. But the furnace cost is high, the power

Y. Chen, *High Temperature Shock Technology*, Nanostructure Science and Technology,
https://doi.org/10.1007/978-981-19-8124-1_5

consumption is large, the workpiece heating speed is slow, the workpiece is easy to oxidize and decarburize when heated in an unprotected atmosphere. Salt bath furnace is the use of molten salt as heating medium furnace type. Although the heating speed is fast and uniform, the oxidation and decarbonization are less, there are parts binding, clamping, and other processes, making the operation complex, labor intensity large, working conditions poor, while there are shortcomings such as long heating time when starting. Induction heating equipment overcomes the above shortcomings. It is easy to use and has good product quality and fast heating speed, but the equipment is more complex, the cost of investment is relatively high. In addition, the interchangeability and adaptability of induction components are poor, so it is not suitable for the application of some complicated workpiece shapes.

HTS is a fast, efficient, environmentally friendly, universal, low-cost, and quantifiable in situ synthesis method, which cannot only overcome the shortcomings of traditional heating in the heating process but also prepare a broad range of materials, such as particles, films, and fibers. Meanwhile, through the design and regulation of HTS equipment and parameters, many kinds of nanomaterials with outstanding physical and chemical performances can be prepared, containing thermal conductivity, flexibility, electrical conductivity, thermal stability, mechanical strength, etc., which can be used as functional devices in various fields, especially in the field of energy storage and conversion [1].

Based on the process advantages and product characteristics of HTS, this chapter will introduce the application of HTS in material preparation equipment and new devices and put forward prospects for further expansion to industrialization and commercialization.

5.2 New Device and Preparation of Nanomaterials Based on HTS

5.2.1 Preparation of Nanoparticles by HTS Combined with Spray Pyrolysis

Fast and scalable fabrication of nanomaterials is the key of widespread use but leaves a challenge. Aerosol spray pyrolysis is a powerful "droplet to particle" nanocrystalline manufacturing approach, which uses a precursor based on solution atomization to make aerosol flow continuously through the heating zone to produce nanoparticles in different states [2–6]. Aerosol spray pyrolysis was used to synthesize various compounds [7], metal oxides [8], and carbon-based materials [9], which have the advantages of simple operation, high yield, low cost, and great scalability, and can prepare materials with different porosity, morphology, stoichiometry, and particle size [9, 10]. However, the current aerosol spray pyrolysis technology usually relies on the tube furnace as the heating element, which can usually reach a temperature of only <1500 K. Moreover, the nanomaterials prepared are often large in size, with low heating efficiency and poor control of quality [11–13].

With the development of spray pyrolysis technology, more and more researchers begin to use this technology to synthesize high-entropy particles, that is, to combine a variety of metals into a single particle at the nanometer scale and form multi-component metal nanoparticles with particular physical and chemical properties [14, 15]. Electronic interactions between different metal components make them widely used materials in the fields of catalysis, plasma, nanomedicine, and electronics [16]. Among all the multi-component metal nanostructures studied, the interest in high-entropy alloy (HEA) nanoparticles has increased significantly. HEA represents a nanostructure defined as five or more nearly equal molar ratio mixed metallic elements [17]. The maximum interaction between different metal atoms forms a unique structure with high entropy and therefore has excellent performances, such as high mechanical strength at high oxidation resistance, high corrosion resistance, and high temperature [18].

So far, the main technologies utilized to manufacture multi-component metal nanoparticles, both in alloy and phase separation states, have been based on wet chemistry, including batch solution synthesis [19], microfluidics [20], and microemulsions [21]. However, the formation of metal nanostructures consisting of more than three metal elements utilizing traditional wet chemistry is extremely challenging as it becomes increasingly difficult to balance the different reduction kinetics of different metal precursors. This usually leads to site-selective nucleation of each metal, as well as the generation of nanoparticles whose composition and structure vary from particle to particle. The melting process is to melt and quench to physically mix various metals to form large size and multi-component HEA [22]. However, recent studies on the atomic distribution of $Al_{1.3}CoCrCuFeNi$ alloys prepared by melting technique have shown that copper-rich precipitates are generated in the final products [23]. In addition, although fusion processing has successfully synthesized a large number of HEAs as structural materials, synthesis at the nanoscale is still difficult. Therefore, there is an urgent need for a technology to compensate for this disadvantage.

1. Microchannel reactor

In 2020, Wang X. et al. proposed a "droplet to particle" aerosol spray synthesis approach, which combined aerosol generator with high temperature (2000 K) microchannel reactor to achieve uniform and fast nanomaterials manufacturing, breaking through the above restrictions. Using this method, unsubstrate materials can be collected for large-scale high-temperature nanomaterials fabrication. The novel developed successive 2000 K aerosol synthesis technology has the following merits:

(1) Each microchannel is 200 mm in diameter and 4 mm in length, 100 times smaller than the size of a representative tube furnace;
(2) This compact size makes heating more uniform and only tens of milliseconds of continuous reduction time;

(3) The microchannel reactor is triggered by electric heating and can reach the temperature of 2000 K easily and efficiently;
(4) The size of the synthesized nanoparticles can be adjusted by the dynamic control of the procedure, including temperature and precursor concentration;
(5) Superior heating capacity (high temperature, tens of milliseconds of duration, and micron size channel uniform heating).

These advantages enable the continuous synthesis of nanomaterials that meet the process requirements, such as high-entropy alloy (HEA) and high-entropy oxide (HEO) nanoparticles. Therefore, the carbonized wood microchannel reactor might be a desired choice for the synthesis of nanomaterials [24].

As shown in Fig. 5.1a, an impact sprayer was used to produce a droplet flow of aerosol gel with a diameter of 1 mm. The metallic salt precursors in the atomized solution are evenly mixed in the resulting droplets, which are then pushed through the microchannels of the carbonized wood reactor by argon gas. When an electric current is applied to the reactor, they are almost immediately and efficiently heated to 2000 K because of the small volume of the droplets inside them and the limited size of the wood microchannels. High temperature drives the decomposition of metal salt precursor to form homogeneous solid solution phase nanoparticles as shown in Fig. 5.1b. Using this method, uniformly mixed HEA PdRuFeNiCuIr nanoparticles can be continuously synthesized, which was confirmed by high-angle ring dark-field scanning transmission electron microscopy (HAADF-STEM) and energy dispersive X-ray spectroscopy (HAADF-EDS) mapping. Figure 5.1c shows uniformly mixed HEA PdRuFeNiCuIr nanoparticles.

Balsa is one of the lightest trees in the world, with a void volume of about 73%, and the individual container channels within are visible to the naked eye, which are carbonized for use as microchannel reactors and offer desired geometry for aerosol droplets. These oval vessel channels, as shown in Fig. 5.2b, c, have an average effective diameter of 200 μm, 200 times the droplet size (1 μm) of the generated aerosol, and 100 times smaller than conventional tube furnaces. Wood perpendicular to the growth direction of the tree was cut to different lengths in order to adjust the residence time of aerosol droplets in the wood microchannel from a few milliseconds to tens of milliseconds. In the process of nanoparticle synthesis, a constant power is utilized to the carbonized wood to offer a joule heating temperature of 2000 K, as exhibited in Fig. 5.2d. Figure 5.2e shows the temperature distribution of the wood. In addition, the temperature can be controlled by adjusting the power applied. As shown in Fig. 5.2f, when the power applied exceeds 600 W, the temperature of the wood can reach 2000 K.

In addition to the higher temperature upper limit, the ability of wood microchannel microfurnaces to significantly reduce holding time is also significantly superior to traditional tube furnaces. At the same flow rate, aerosol droplets can be held for 2 ~ 3 orders of magnitude longer in conventional furnaces than in wood microchannels. For example, the insulation time of aerosols in a 40 cm conventional furnace was 1.6 s, while it was only 16 ms in a 4 mm wood microchannel. In addition, prolonged heating is likely to result in further growth of nanoparticles. The reaction rate will be

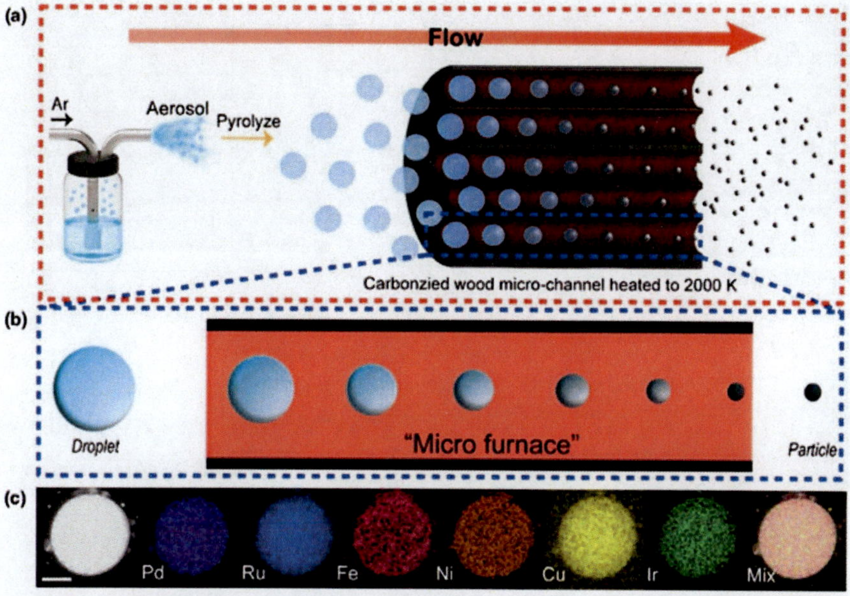

Fig. 5.1 **a** Schematic diagram of droplet transport via carbonized wood microchannels heated by carrier gas to 2000 K; **b** enlarged diagrams of droplet to particle evolution in single wood microchannels; **c** HAADF-STEM images and HAADF-EDS images of PdRuFeNeCuIr nanoparticles

greatly increased if the temperature restriction of the microchannel reactor is higher. Hence, the reaction can be completed in a shorter holding time, while also preventing particle sintering because of less accumulated heat.

In addition, due to the significant reduction in the size of the carbonized wood, the heating efficiency is much higher than that of traditional furnaces. The comparison of the sizes of traditional tube furnace and carbonized wood microchannels is shown in Fig. 5.2g, h. The size of the wood microchannels results in more efficient and even heating. To verify this, the temperature distribution along the tube furnace and the carbonized wood microchannel was simulated, respectively. The temperature distribution simulation showed that the temperature of the carbonized wood microchannel could rise rapidly to 2000 K, while the temperature rise was much slower for the traditional furnace. Selected simulations along a conventional furnace cross section and temperature profiles with lengths of ¼, ½, 1, and 10 times the tube diameter are exhibited in Fig. 5.2i, j. In Fig. 5.2i, 0.5 cm (for ¼ conventional furnace diameter) has a center temperature of only 350 K. Even 20 cm (10 times furnace diameter) has a center temperature of merely 900 K. Nevertheless, in Fig. 5.2j, the microchannel length is 50 mm (equal to ¼ diameter), and the temperature is up to 1400 K. The length of the microchannel is 200 mm (equal to the diameter), and the temperature reaches 2000 K. The carrier argon is heated to 2000 K in 0.1 ms, and the aerosol droplets are heated to 2000 K in 1 ms. The heating speed of argon is as

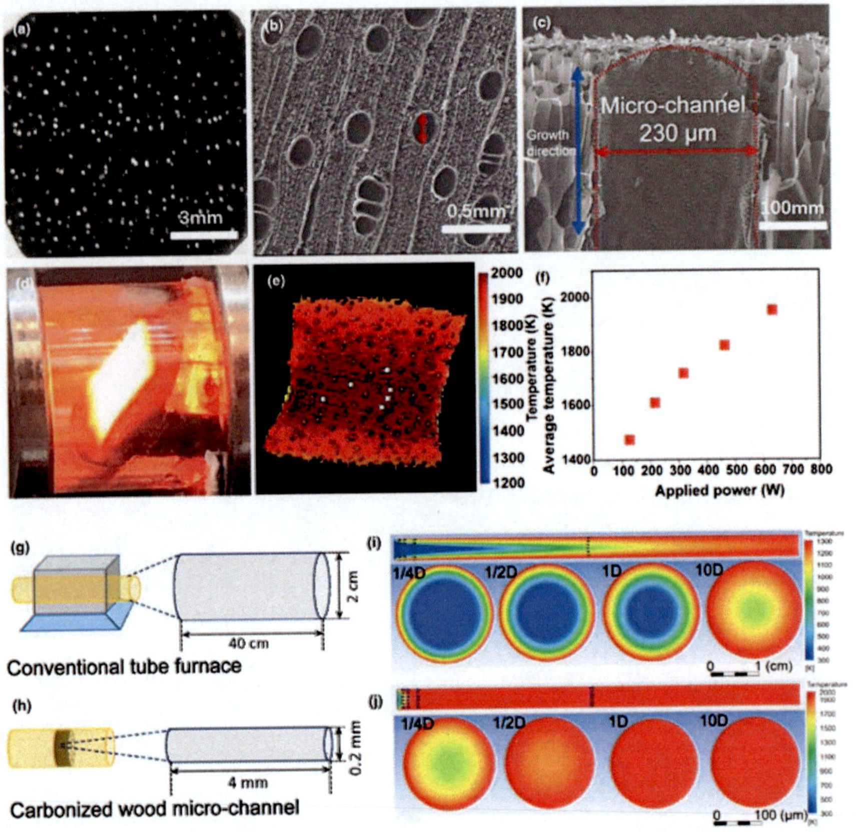

Fig. 5.2 **a** Photograph of carbonized balsa wood in which bright spots come from lines of light passing through open channels; **b** SEM image of top view of wood microchannels, looking down at the length of tree growth direction; **c** SEM image of cross section of wooden microchannel; **d** photographs of carbonized wood heated to 2000 K; **e** temperature measurement of carbonized wood heated to 2000 K; **f** average temperature of carbonized wood and applied power; **g** schematic diagram of a conventional tube furnace; **h** schematic diagram of microchannels in carbonized wood; **i** the temperature distribution along the length and cross section of a 1300 K conventional tube furnace with lengths equals to ¼, ½, 1, and 10 times the tube diameter, respectively; **j** the temperature distribution of carbonized wood microchannels along lengths equals to ¼, ½, 1, and 10 times the diameter of wood microchannels, respectively

high as 10^7 K/s, while the heating speed of aerosol is 10^5 ~ 10^6 K/s. This is because hot argon accelerates the heating of aerosols because there is a positive temperature gradient between them. In contrast, for conventional furnaces, while aerosols appear to be heated to 1300 K in 0.01 s, argon is heated much more slowly because the negative temperature gradient between them results in reduced heating efficiency of aerosol particles and a slower increase in the actual temperature of aerosols.

Compared with traditional tube furnaces, microchannel reactors exhibit some different characteristics, as shown in Table 5.1.

Table 5.1 Comparison between microchannel reactor and traditional tube furnace

Parameter	Traditional tube furnace	Microchannel reactor
Channel diameter	≥2 cm	≈200 μm
The length of the channel	≥40 cm	Adjustable height, ≈mm
The temperature distribution	Non-uniform	Uniform
The temperature limit	<1500 K	≥2000 K
The reaction rate	Low	High
The duration of the	>1 s	≈10 n ms
Ability to synthesize HEA	Weak	Strong

Besides balsa, other kinds of carbon-based materials can be utilized as microreactors for successive synthesis, containing different woods (e.g., poplar with 100 mm of channel diameter), wood drilled artificially to achieve desired aperture and quantity, and highly crystalline 3D-printed carbon frames such as reduced graphene oxide (RGO).

If the precursors can be atomized into droplets, various materials can be prepared through carbonized wood reactors. Wang et al. synthesized and characterized HEA and HEO nanoparticles, which can be obtained by fast heating and cooling at high temperatures. Elemental analysis showed that the different elements were uniformly dispersed in the particles, and no obvious element separation or phase separation was observed. For comparison, we atomized the same precursor and passed it through a conventional tube furnace at 1300 K. Nevertheless, utilizing this approach, the particles show serious phase separation. This observation suggests that HEA nanoparticles can be synthesized by the superior heating capacity of wood microchannel reactor.

The technique can also be extended to prepare more complex HEA nanoparticles, such as those consisting of 10 different elements (Au, Co, Cu, Ir, Pt, Mn, Mo, Ni, Pd, Ru). These elements have important differences in physical and chemical properties, such as atomic radius, melting and boiling point, crystal structure, and metal precursor properties, showing different chemical reduction potentials and physical decomposition temperatures. In spite of these challenges, the combination of aerosol spray and our microchannel reactor was able to produce complex HEA nanoparticles, and elemental analysis showed that the 10 elements were evenly distributed.

A detailed diagram of droplet to particle synthesis achieved by the carbonized wood microchannel reactor is shown in Fig. 5.3a. After the precursor solution is atomized into droplets, they are transported by carrier gas to the natural channels of the carbonized wood and heated to 2000 K to be thermally decomposed into particles. As with conventional aerosol synthesis, as the droplet flows through the microchannel, the remaining solvent in the droplet evaporates fast upon heating, so that each droplet evolves into a uniform and dense metallic salt particle. In addition, since 2000 K is above the melting point but below the boiling point of most metals, it is reasonable to assume that in each droplet, metallic elements tend to mix in the liquid phase to form liquid metal alloys. Considering that the holding time of the carbonized wood microchannel is merely 16 ms, the liquid metal alloy particles are fast quenched to indoor temperature after leaving the reactor to maintain the uniform mixing state of elements, and HEA particles are collected at indoor temperature.

Figure 5.3b shows how temperature affects the nucleation of CoNiPdRuIr HEA nanoparticles. When the temperature of the carbonized wood microchannel reactor was 1400 K, the temperature was too low to allow the complete pyrolysis of the droplet, resulting in the formation of irregularly shaped particles. In addition, phase separation on the uneven surface of these materials because of low temperature can be clearly observed, probably because merely some precious metal salts (Pd, Ru, Ir) with low decomposition temperature can fully decompose and nucleate, generating the

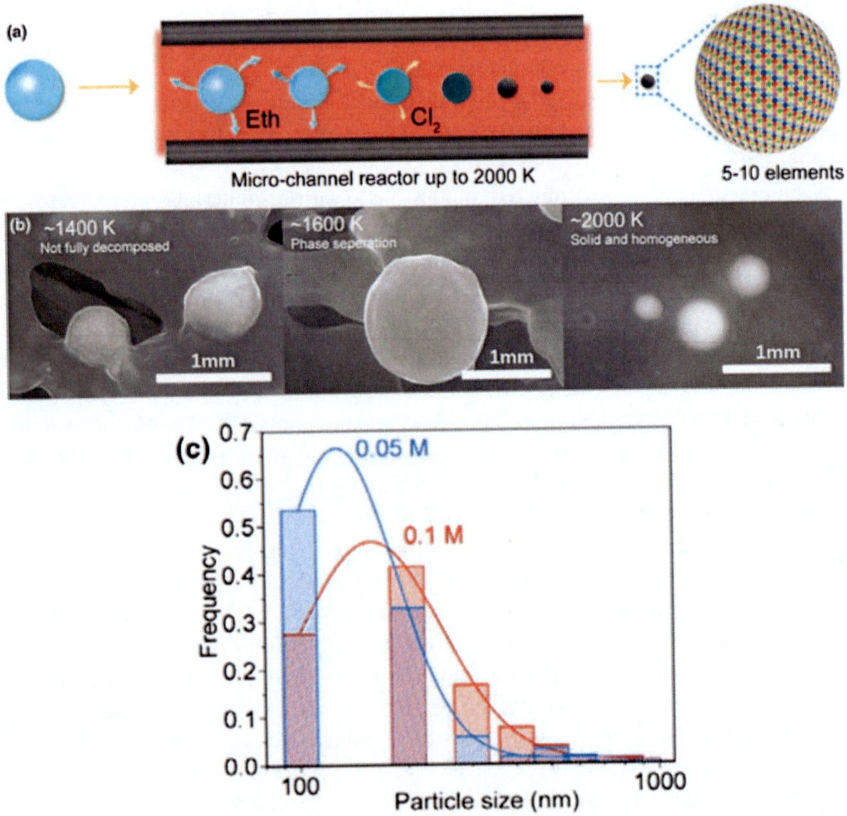

Fig. 5.3 **a** HEA formation mechanism in the carbonized wood microchannel reactor; **b** SEM images of CoNiPdRuIr HEA particles synthesized at different temperatures; **c** size distribution of CoNiPdRuIr HEA nanoparticles synthesized at different precursor concentrations (0.05 mol/L and 0.1 mol/L)

bright spots exhibited on the surface of these materials in Fig. 5.3b. When the temperature is raised to 1600 K, regular round particles are generated. Nevertheless, phase separation can still be seen on the surface. In contrast, when the temperature is high enough (i.e., 2000 K), the decomposition speed is obviously increased, and the metal salts can decompose completely. In addition, uniform liquid alloys can be formed merely at sufficiently high temperatures (2000 K). After fast quenching, uniform high-entropy mixture is obtained. Figure 5.3c shows the size distribution of HEA nanoparticles synthesized with precursors of different concentrations (0.1 mol/L and 0.05 mol/L). The results exhibited that the particle size increased with the increase of precursor concentration, and the average diameter of particles with high concentration (0.1 mol/L) was 194 nm larger than that of particles with low concentration (0.05 mol/L), while the average diameter of particles with low concentration was merely 147 nm. These findings suggest that particle size can be dynamically adjusted.

Aerosol spray pyrolysis has the advantages of low cost, simplicity, and low waste emissions and is very suitable for successive large-scale preparation of high-quality HEA nanoparticles.

2. Tubular reactor

In 2019, Yong Yang et al. developed a carbon thermal shock (CTS) technology that combines multiple non-miscible metallic elements into a single nanoparticle. This technology can alloy up to eight metallic elements into single-phase nanoparticles evenly distributed on a carbon scaffold. It mainly consists of ultra-fast electric heating, thermal decomposition of carbon matrix metal salt mixture into liquid metal particles, and subsequent ultra-fast quenching steps to prepare solid HEA nanoparticles with well-mixed metal elements.

Though aerosol synthesis techniques have been extensively researched for the preparation of metals [25], alloys [26], metal oxides [27], and composite metals [28], Yong Yang et al. demonstrated the ability to produce nominally unmixable materials by mass-producing HEA powders without predepositing the metal salt mixture precursors onto carbon carriers. Instead, the precursors were decomposed into aerosol droplets less than 1 μm in diameter and treated with rapid heating and quenching. As a result, each droplet containing a precursor acts as a nanoreactor with a very low thermal mass, allowing fast heating and cooling to produce a single HEA nanoparticle. Yong Yang et al. demonstrated the aerosol-mediated synthesis of five-element HEA nanoparticles containing Ni, Co, Cu, Fe, and Pt, which have extensive applications in the fields of catalysis chemistry, magnetism, and electronics [29]. A combination of atomic resolution scanning transmission electron microscopy (STEM) and energy dispersive X-ray spectroscopy (EDS) analysis showed that all metallic elements (incompatible metals such as Co and Cu) in the nanoparticles were mixed at the atomic scale. The advantages of utilizing aerosol droplet mediated technology are as follows:

(1) All metal salts are confined to small droplets, and different metals can be added during heating and quenching;

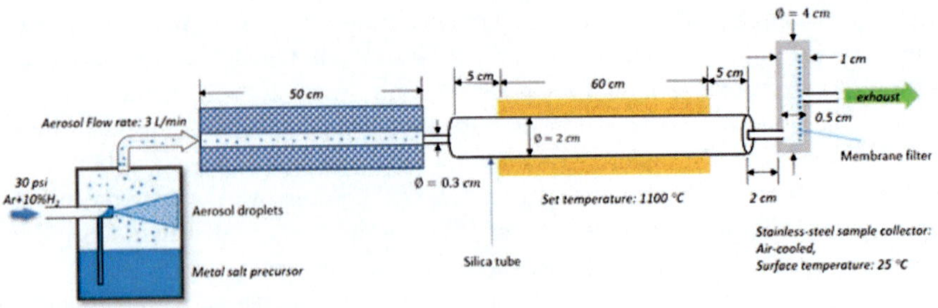

Fig. 5.4 Schematic diagram of aerosol reactor, detailed geometry, and operating parameters

(2) The small mass and volume properties of aerosol droplets can be heated and cooled rapidly, which is very significant for the dynamic control of thermodynamic mixing system and the generation of HEA nanoparticles;
(3) The metal salt ratio in aerosol droplets reflects the metal ratio in precursor solution, so that the composition and composition of the final product can be adjusted only by adjusting the kind and proportion of metal salt in precursor solution;
(4) The technology is a successive production procedure.

Figure 5.4 is a schematic diagram of the aerosol reactor, containing geometry, flow rate, operating pressure, and temperature distribution. The overall length of the tubular reactor (silicon tube, inner diameter 2 cm) is approximately 70 cm, with a region 60 cm long in the middle, heated by two combined tube furnaces. There is an area 5 cm long at each end of the tube that is not heated by a heating coil. The heating furnace temperature is set at 1100 °C. The measured gas flow speed is about 3 L/min, and the residence time of aerosol droplets through the heating zone is about 3 s.

Figure 5.5a is a schematic diagram of the aerosol droplet mediated scalable synthesis of HEA nanoparticles. The precursor solution including five kinds of metal salts in deionized water was atomized, and aerosol droplets were generated by using a colloidal atomizer. The gas utilized to fog the solution is a 10%H_2/90%Ar mixture to promote the generation of zero-valent HEA. Carried by the mixture of droplets atomized aerosols by silicon dryer removing solvent water, then, made up of five kinds of mixed metal salt dry aerosol particles through high temperature reaction zone (about 1100 °C), the temperature rising fast (thermal shock) trigger metal salt elemental fast pyrolysis and reduction of metals, resulting in rapid HEA nanoparticles is formed after quenching. Particles leaving the reactor are collected on a membrane filter collector.

Figure 5.5b shows the distribution of elements within a single NiCoCuFePt HEA nanoparticles with a diameter of 60 nm. Ni, Co, Cu, Fe, and Pt elements were observed to be uniformly distributed in the nanoparticles. The five elements exhibited in the EDS diagram overlap with each other, indicating the successful synthesis of HEA nanoparticles by aerosol droplet mediated method. X-ray diffraction (XRD)

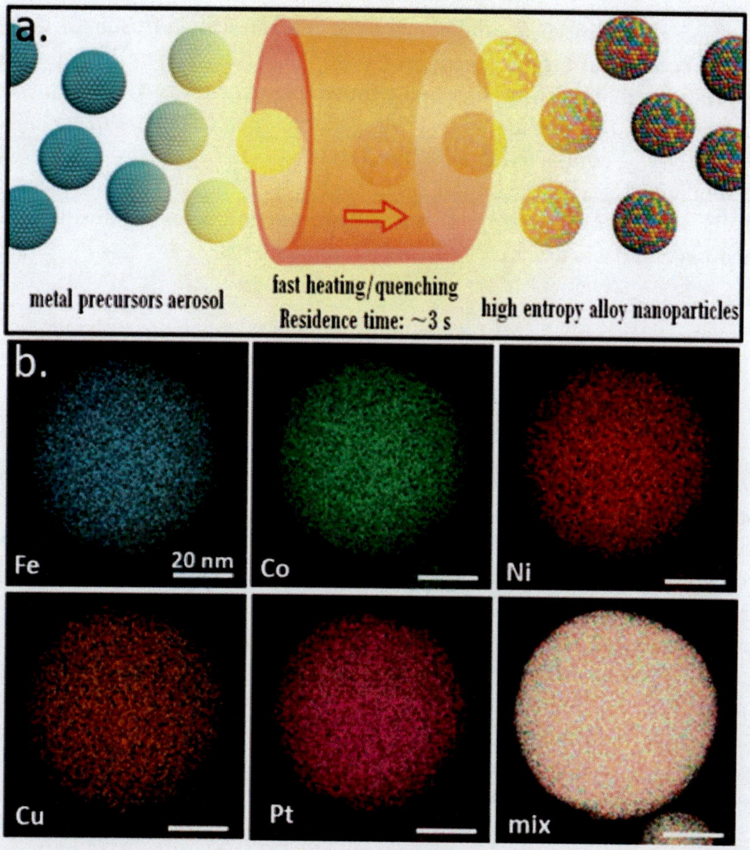

Fig. 5.5 Aerosol droplet mediated scalable synthesis of HEA nanoparticles **a** Schematic diagram of aerosol droplet evolution during high temperature treatment (the actual particle size is much smaller than the tube size exhibited in the figure); **b** EDS spectrum of single HEA nanoparticles

shows that the single-phase structure and EDS data of synthetic particles are further confirmed.

This aerosol-assisted procedure provides a general method to synthesis. Because most of the metal elements in this research have relatively low standard potentials, heat is request to reduce the metal salt precursors produced by hydrogen. Hydrogen reduction of metallic nitrate precursor can be written as $M(NO_3)_2 + 2H_2 \rightarrow M + 2NO_2\uparrow + 2H_2O\uparrow$.

The conversion of metal salt precursors to metal particles involves two procedures: diffusion of hydrogen molecules into aerosol particles and reduction of metal salt precursors by hydrogen molecules. Since particles are kept in the high temperature reaction zone for a very short time (about 3 s), it is necessary to consider whether hydrogen molecules are dynamically advantageous. Earlier researches have exhibited that hydrogen molecules can diffuse through gaps in metals. Because the diameter of the particle (<10^3 nm) is far smaller than the characteristic diffusion distance and diffusion procedure at high temperature (1100 °C), so hydrogen diffusion should not be rate-limited. Therefore, the high temperature accelerates the kinetics of the reduction of metal salt precursors by hydrogen, and the conversion of metal salt to metal phase can be completed in less than 3 s.

The simple and flexible technique of aerosol droplet mediated by tubular reactor extends the method of preparing HEA nanoparticles and demonstrates its potential to extend beyond basic research to industrial applications.

5.2.2 *Preparation of Nanoparticles by HTS Combined with Flexible Materials*

Supported nanoparticles are widely utilized in plasma, catalysis, energy storage, antibacterial, and energy conversion fields [30–33]. Traditional high-temperature approaches such as annealing, thermal reduction, and calcination provide a good thermodynamic environment for efficient and controllable synthesis of nanoparticles [34, 35]. In addition, shorter duration methods, such as microwave radiation, can generate nanoparticles in minutes or even seconds [36]. These physical deposition approaches based on high temperature procedures are easy to scale up and sustainable, with little chemical contaminants and solution waste and are conducive to the synthesis of loaded nanoparticles.

Nevertheless, these approaches heat the whole sample to high temperatures and are hardly suitable for substrates with low thermal stability (e.g., textiles and paper) whose structures are vulnerable to damage during heating. Nanoparticles dispersed on flexible and lightweight low-temperature substrates are increasingly important for a variety of medical and wearable applications [37]. These medical and wearable applications need a simple and scalable method of particle synthesis and should be compatible with substrates with low thermal stability [38, 39]. Hence, it is necessary to break through the limitation between efficient high-temperature synthesis and sensitive substrates with low thermal stability to rapidly synthesize nanoparticles in a scalable and sustainable manner.

In 2019, Miaolun Jiao et al. reported a method for the synthesis of well-dispersed nanoparticles on a low thermal stability matrix through high temperature and rapid radiation heating. The textile loaded with the metal precursor passed through the top of the heat source (2000 K) at a fixed rate of 0.5 cm/s (distance 0.5 cm) and the high temperature source nucleated the precursor salt and nanoparticles. In addition, continuous synthesis of nanoparticles under limited short heating time can also avoid deterioration of the textile substrate. This non-destructive method is suitable for the synthesis of metal nanoparticles on a variety of low thermal stability substrates, and this non-contact and successive synthesis is compatible with the roll-to-roll process of ultra-fast nanomaterials [40].

The rapid radiation heating procedure is shown in Fig. 5.6a. The textile loaded with H_2PtCl_6 precursor passes through the high temperature heat source at a successive speed of 0.5 cm/s, and the formation of nanoparticles is induced by high temperature radiation and subsequent quenching. The high intensity of thermal radiation caused the decomposition of the H_2PtCl_6 precursor and its conversion to Pt nanoparticles on the textile surface within seconds. Due to the short duration of the high temperature treatment procedure, the structure of the textile remains basically unchanged, avoiding the significant degradation caused by high temperature. Therefore, the radiation heating method is particularly suitable for the synthesis of metal nanoparticles on textiles with low thermal stability and is similar to the traditional long-term heating method, which is beneficial to improve the production efficiency.

In this method, to load the precursor salt, textiles are soaked in H_2PtCl_6 precursor solution (0.05 mol L^{-1} H_2PtCl_6 in ethanol) for 20 min and then dried in air for 1 h. Figure 5.6b shows crystal salts of 200 ~ 400 nm sparsely dispersed on the surface of the textile fiber. In contrast, many platinum nanoparticles are uniformly distributed on the fabric with particle sizes between 60 and 140 nm after the radiation heating process, as shown in Fig. 5.6c. At the same time, after the rapid radiation heating method, the surface of the fiber is largely undisturbed and its structure remains. EDS as shown in Fig. 5.6d further verifies the decomposition of H_2PtCl_6

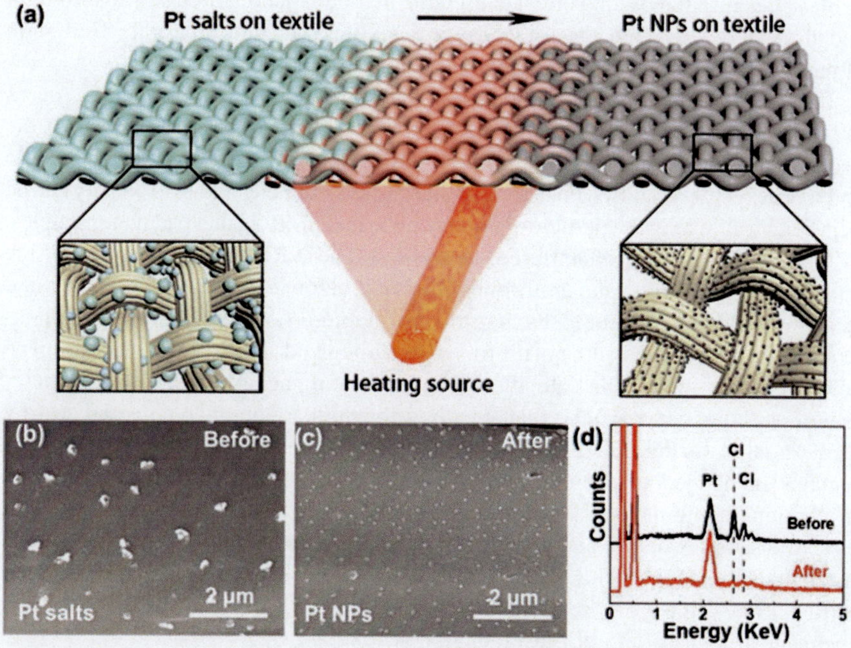

Fig. 5.6 **a** "Fly through" diagram of platinum nanoparticles loaded on textiles synthesized by fast radiation heating; **b**, **c** morphology of H_2PtCl_6 precursor loaded on textiles before and after radiation heating; **d** EDS spectra of H_2PtCl_6 precursor before and after radiation heating

precursor. In the samples treated with radiation heating, the peak value of chloride was significantly reduced, indicating that most of the chlorine elements disappeared after thermal decomposition, leaving metallic Pt nanoparticles. Thus, through rapid radiation heating, precursor salts can be decomposed into metal nanoparticles in a short time without major changes to the textile fibers.

Figure 5.7 describes the tunable parameters of the radiation heating approach for the preparation of loaded nanoparticles. Figure 5.7a illustrates the setup and processing parameters of the rapid radiant heating approach, including the temperature of the radiant source (Ts), the working distance between the heat source and the low-temperature substrate (d), and the velocity of movement/production (Vp). In the experiment, carbon nanofiber (CNF) films were utilized as a radiation heating source, which could be heated up to 3000 K by electric joule heating, and the heating was controlled by adjusting the electrical input. Figure 5.7b shows textiles loaded with the H_2PtCl_6 precursor being pulled across the radiant heating zone at a velocity of 0.5 cm s^{-1} at 0.5 cm from the emitter. After the heat source, the surface of the textile is obviously black, while the rest of the textile is still the original white. Control tests on raw textiles without salt-free precursors showed no change in color or form after radiation heating, as shown in Fig. 5.7c, manifesting that fast heating does not cause obvious decomposition or carbonization of textile substrates. The color of the textiles loaded with the precursors changed from yellow (as opposed to H_2PtCl_6 precursors) to black because the plasma effect caused the platinum nanoparticles to absorb light and turn black. Therefore, the color change of textiles is an indicator of the formation of nanoparticles during radiation heating.

If the salt precursor is decomposed completely, the rapid radiation heating process presents a similar size distribution. As a result, the processing parameters in the heating procedure can be easily adjusted to similar nanoparticles, proving its flexible and robust synthesis ability. For example, the particle diameter and load density can be adjusted by adjusting the particle diameter and load density of the load nanoparticles.

Optical microscopy, FT-infrared, tensile test, and XRD were used to study the morphology, durability, and uniformity of textiles before and after radiation heating, as shown in Fig. 5.8. Figure 5.8a, b is an optical microscope image, further demonstrating that the textiles were not destroyed or carbonized after radiation heating. The color of brown of the substrate after radiation heating again manifests the generation of platinum nanoparticles that absorb visible light, rather than potential destroy to the textile. In Fig. 5.8c, the infrared display of O–H, C–H, C = O, and C–O remains unchanged, proving that carbonation is negligible during radiation heating. In addition, titanium OlsenH5KT tester was used to measure the mechanical strength of textile samples before and after radiation heating, as exhibited in Fig. 5.8d. The strength of Pt nanoparticles loaded textiles (45.9 MPa) was like that of the original textiles (50.7 MPa). The results showed that the textile retained its physical and chemical structure after ultra-fast radiation heating.

Structural integrity after rapid synthesis of low thermal stability substrates is one of the most promising aspects of this method and can be easily extended to other substrates. The substrate used in these experiments is a loose woven fabric that can better load the precursor salt without any additives on the surface of the substrate.

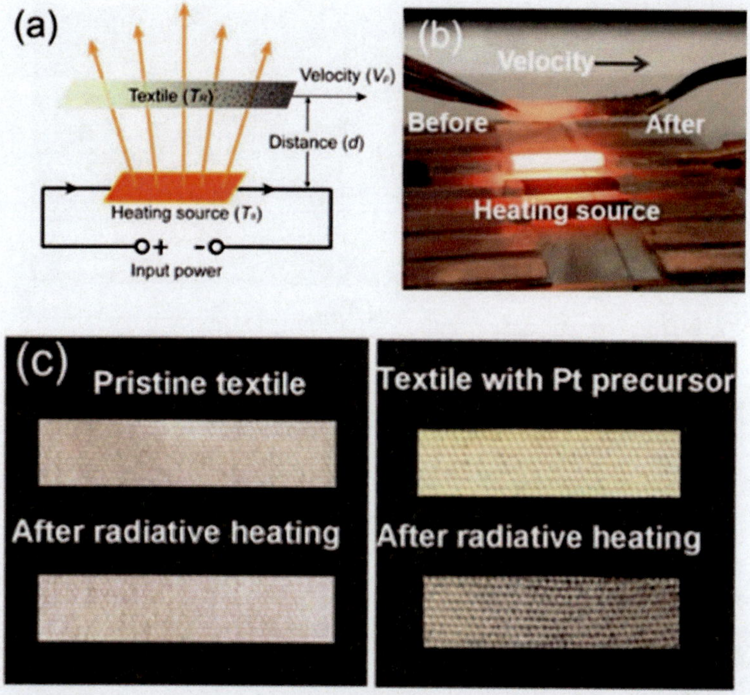

Fig. 5.7 Dynamic control during rapid radiation heating **a** Side-view sketches detailing the parameters of the radiant heating approach; **b** Image of the radiant heating process of pulling the prebody loaded fabric through the corresponding side view of the radiant heating source. **c** Before and after photographs of the original textiles and precursors loaded textiles, respectively, subjected to radiation heating

Therefore, this rapid radiation heating method is a promising method for the synthesis of uniform nanoparticles on flexible porous substrates with low thermal stability.

Figure 5.9 is a diagram of an ultra-fast radiative heating process for roll-to-roll manufacturing based on low thermal stability. The low-temperature substrate is immersed in the precursor salt solution, and the metal nanoparticles are rapidly synthesized by a radiation heating source. This method is compatible with existing industrial protocols and has a good prospect in the field of ultra-fast roll-to-roll nanomaterials with various loaded nanoparticles.

5.2.3 *Preparation of Nanoparticles Based on Micro-reactor of HTS*

High temperature is one of the most common and powerful approaches of fabrication for synthesizing various materials in the laboratory and industry, including metals,

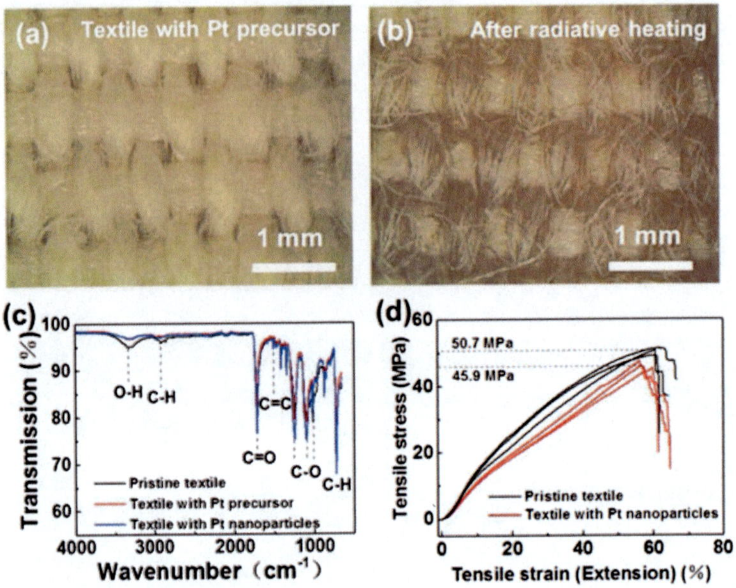

Fig. 5.8 Morphology and structural evolution before and after radiation heating **a**, **b** Microscopic images of textiles before and after radiation heating, where the blackness is because of the synthesis of nanoparticles on the fabric after radiation heating; **c** FT-IR of raw textiles loaded Pt precursors and radiation-heated textiles; **d** mechanical strength of Pt precursor textiles before and after radiation heating

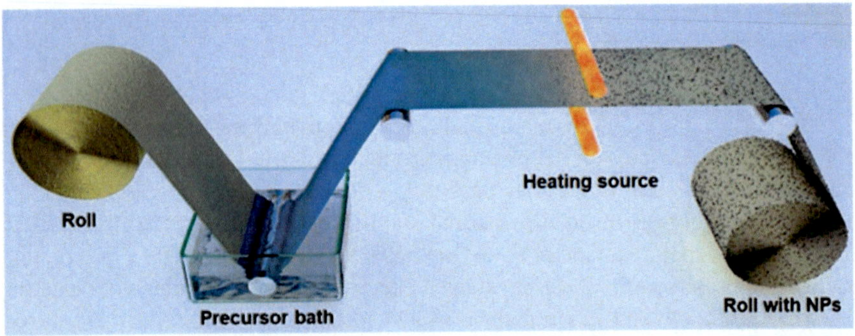

Fig. 5.9 Schematic diagram of the preparation of nanoparticles for the roll to be used in low-temperature basis

ceramics, and carbides [41, 42]. However, traditional sintering furnaces are based on radiation heating in a large chamber and usually have a limited temperature range (<1500 K), slow heating/cooling rate (<100 K min^{-1}), low energy efficiency (<10%), which will also result in large temperature gradient and uneven heat distribution in bulk material processing, leading to uneven structure of synthetic materials [43–47]. Prolonged heating at high temperatures impedes the application of rapid heat treatment and reaction in non-equilibrium processes, especially for the synthesis of nanomaterials, which requires highly controlled and delicate heating conditions to achieve particularly designed surfaces and nanostructures.

In 2020, Yun Qiao et al. developed a versatile and highly controllable microreactor for in situ synthesis of homogeneous nanoparticles on a carbon carrier utilizing electric joule heating, which will facilitate rapid synthesis of high-temperature annealing-sensitive nanostructures [48].

In this work, they demonstrated that a 3D-printed small reactor can be used for batch and scalable synthesis of nanocatalysts through efficient and controlled heating. The reactor itself is heated by direct joule (>90% conversion efficiency) trigger, reaching a temperature of $\approx$3000 K, with an ultra-fast heating speed of $\approx 10^4$ K^{-1}. As a proof of concept, they manufactured a small reactor utilizing a 3D extrusion print-based graphene oxide (GO) slurry, first printing a dense layer of GO as the bottom, then using six layers of extrusion wire to construct a microscale grid pattern, as shown in Fig. 5.10a. The resulting 3D printing microreactor (20 mm $\times$ 10 mm $\times$ 2.2 mm) has a matrix structure and offers space for storing raw materials for subsequent reactions. Then, GO was converted to RGO with medium conductivity in a thermal reduction reactor at 300 °C in argon atmosphere for 1 h to generate high temperature through joule heating. Figure 5.10c shows a 3D-printed photo with a temperature up to $\approx$3000 K. The reactor is characterized by a dense grid pattern that maximizes efficient and uniform thermal conductivity of the material, which is critical for achieving highly controlled heating throughout the treated material, as shown in Fig. 5.10d. In addition, the shape, size, and internal structure of the reactor can be easily adjusted by a 3D printing program. Small chambers including each part of the grid 3D printing reactor allow for highly controlled heating to be maximized for efficient and rapid heat treatment. Thus, by combining the GO microunit and joule heating, shape design and highly controlled high temperature heating can be achieved with significantly less energy and cost than conventional furnaces.

In order to explore the heating capacity of the reactor, the device was suspended on two glass sliding sheets and electrically heated by joule with silver paste connected to copper electrodes, as shown in Fig. 5.11a. The photo of the reactor in Fig. 5.11b, c shows different brightness states under different current pulses (10 ~ 20A), manifesting that the temperature can be controlled by the input power supply. In addition, a 4000 frame/s high-speed camera was used to obtain the spatial temperature distribution of the 3D printing reactor, and the time and temperature changes were estimated based on colorimetry, as shown in Fig. 5.11d. Thermal images showed the spatial distribution of temperature in the 3D reactor, where all cavities were uniformly heated at similar temperatures ($\approx$2400 K). This is essential for the uniformity of batch synthesis. Figure 5.11e shows the change in reactor temperature over time. After the first heating $\approx$30 ms, the temperature rises fast to $\approx$2300 K and remains at $\approx$2400 K about 200 ms. At the end of the reaction, the temperature drops from 2400 to 1700 K only in 30 ms, and the calculated heating and cooling rate are 10^4 K/s.

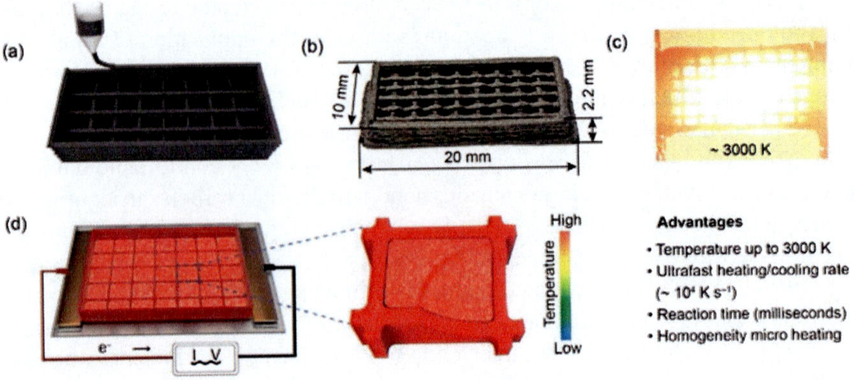

Fig. 5.10 Small cavities in the matrix structure of a 3D-printed small reactor **a** Schematic diagram; **b** 3D-printed photos of the microreactor; **c** 3D-printed photos of the RGO reactor at temperatures up to ≈3000 K; **d** schematic diagram of sample temperature distribution in the 3D-printed microreactor and advantages of the reactor in the joule heating process

This ultra-fast rate is critical for achieving rapid reaction and avoiding overheating resulting in deterioration of material structures such as nanoparticle polymerization.

Figure 5.11f shows the radiation spectrum of a 3D-printed RGO reactor. The input current is collected by optical fiber and sent to the spectrometer. The higher the input current is, the higher the emission intensity and temperature are. In addition, Qiao et al. proved a fast thermal shock with temperatures up to 2000 K, as shown in Fig. 5.11g, where high heating and cooling rates can be clearly observed. Figure 5.11h shows the cycle stability and durability of the 3D printing reactor at 2000 K, where the reactor keeps its radiation heating characteristics without obvious change after ≈300 cycles. These results indicate that 3D-printed small RGO reactors are excellent candidates for high-temperature reactions, with ultra high temperatures and ultra-fast heating and cooling speeds in addition to their stable and durable performance.

Qiao et al. conducted batch synthesis of loaded nanoparticles using a miniature reactor of 3D printing, which proved its practicality. Figure 5.12a shows the loading, fast heating, and sample unloading of the precursor. Figure 5.12b shows a photo of the reactor loaded with the feedstock powder and during heating, where conductive and radiative heating occur within the channel of the substrate, converting the precursor to nanoparticles. After the radiation heating procedure, the powder remains in the reactor, and the reactor itself remains unchanged. The morphology and particle size distribution of nanoparticles were analyzed by TEM as shown in Fig. 5.12c, d. TEM images and particle size distribution confirmed the existence of nanoparticles (average size ≈2 nm), which were uniformly distributed and embedded in the interconnected channels of the structure. For comparison, the precursor salts in the base of a tube furnace were annealed in an argon atmosphere. At a heating speed of 5 K min^{-1}, the temperature was firstly heated to 1073 K for 1 h. The subsequent natural cooling process takes about 4 h to bring to room temperature. The whole process takes 7 h,

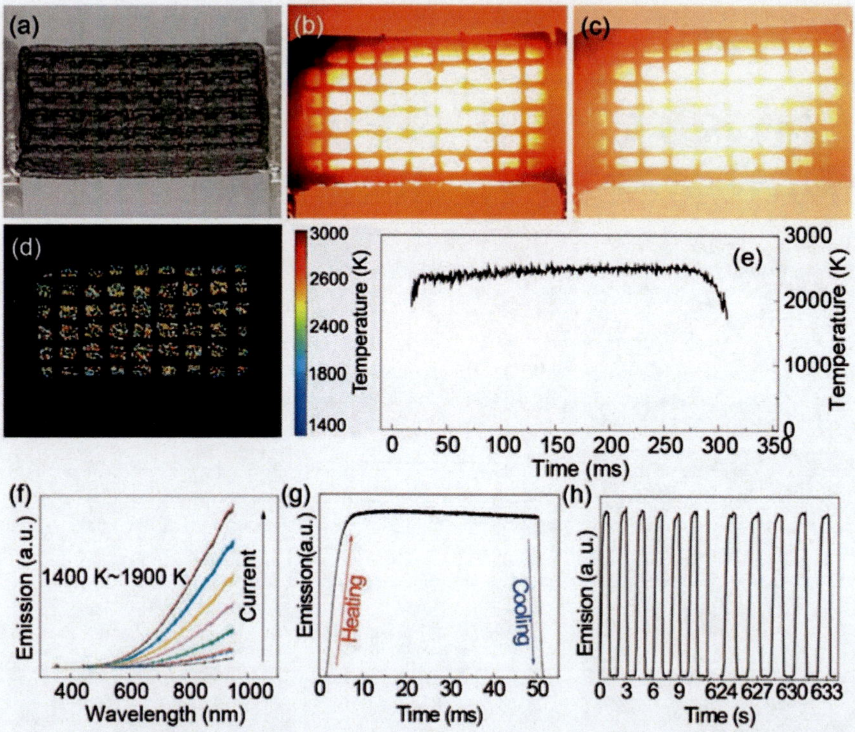

Fig. 5.11 High temperature performance of 3D-printed microreactor **a** 3D-printed miniature RGO reactors; **b**, **c** photographs of the same facility at different temperatures; **d** spatial distribution of temperature in a three-dimensional reactor; **e** change of reactor temperature with time; **f** luminescence spectra of the reactor at different input currents; **g** fast thermal shock behavior of the reactor; **h** cyclic performance of RGO reactor

significantly longer than the reactor (500 ms). TEM images showed that the nanoparticles prepared using a tube furnace were highly concentrated in the substrate and had a wide size distribution due to the long processing time and slow heating and cooling process. Figure 5.12e, h shows high-resolution TEM images of individual nanoparticles prepared utilizing a miniature reactor and a tube furnace using 3D printing, respectively. These results indicate that the rapid heating capability of a small reactor enables fast and efficient synthesis of homogeneous nanomaterials.

In conclusion, this versatile 3D-printed microreactor can offer efficient, rapid, and controllable heating for the synthesis of high-temperature nanocatalysts. It increases material contact during thermal shock through a simple 3D-printed microscale grid design that can reach temperatures up to ≈3000 K, heating times in milliseconds, and rising speeds up to 10^4 K/s. The combination of thermal shock approach and three-dimensional small reactor can be used for fast and batch synthesis of loaded single metal or alloy nanoparticles, which will be widely used in energy storage and catalysis.

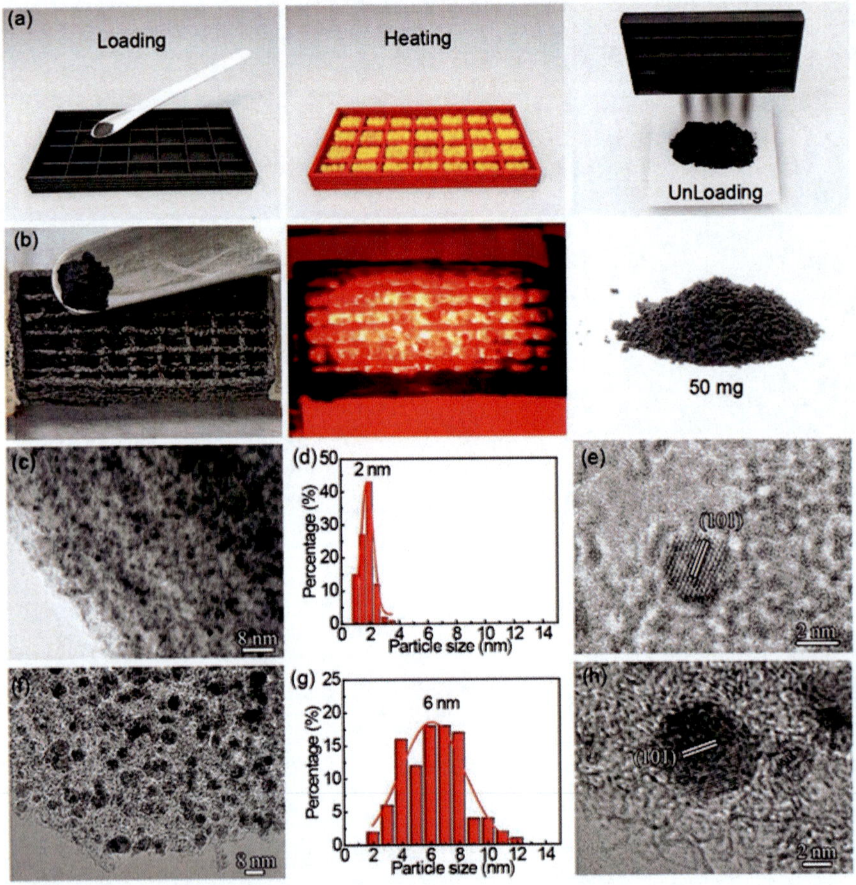

Fig. 5.12 Demonstration of a 3D-printed microreactor for the preparation of ruthenium nanoparticles embedded in CMK-3 **a** Schematic diagram of powder loading, heating, and unloading processes in a miniature reactor using 3D printing; **b** photographs of the reactor during sample loading and heating at 1500 K, and Ru nanoparticles thus embedded in CMK-3; **c** TEM images; **d** corresponding statistical size distribution; **e, f** high-resolution TEM images of Ru nanoparticles in CMK-3 prepared utilizing a 3D-printed microreactor; **g** corresponding statistical size distribution; **h** high-resolution TEM images of Ru nanoparticles in CMK-3 prepared in a tube furnace

5.2.4 Preparation of Self-Supporting Carbon Materials by HTS Combined with Roll-To-Roll Technology

In small electronic devices with dense integrated circuits, the rapidly increasing thermal flux density presents a major challenge to thermal management materials [49, 50]. Commercially available polyimide carbon films produced by sequential carbonization and graphitization have high in-plane thermal conductivity (up to 1750 W/mK), making them eligible for use as heat transfer materials in modern microelectronics and optoelectronic applications. Nevertheless, the serious need for the molecular structure of the original polyimide and the complicated production process (containing polymerization, film making, film making treatment, long-term carbonization, graphitization, etc.) severely reduces the production efficiency of the product. In addition, another insurmountable problem is the enormous energy consumption when preparing high thermal conductivity carbon films using traditional graphitization furnaces. Achieving advanced thermal management materials with optimal performance need scalable and cost-effective manufacturing methods combined with new chemicals as well as synthetic routes.

Graphene, a thin sheet of carbon atoms arranged in a two-dimensional hexagonal lattice, provides great potential for the next generation of thermal management [51–54]. Macro-graphene films assembled from graphene sheets are thought to be ideal heat dissipating materials for high power facilities because of their combined advantages of high thermal conductivity and flexibility. In order to pursue high thermal conductivity films, high temperature annealing in electric arc furnace is indispensable for eliminating oxygen-containing groups and recovering structural defects [55]. As the furnace itself limits the heating/cooling rate, it usually takes at least two days to accomplish the annealing procedure [56].

Graphene fibers, as a new member of carbon fibers assembled from graphene sheets, inherit the superior electrical, mechanical, and thermal performances of inherent graphene [57–59]. The current popular strategy for synthesizing graphene fibers is wet spinning of GO, followed by reduction of GO fibers to remove oxygen-containing functional groups and repair defects. The reduction procedure can improve the electrical, mechanical, and thermal performances of fibers by several to tens of times [56, 60, 61]. Among them, high temperature heat treatment is one of the most effective methods to RGO fibers, eliminates non-carbon impurities, and improves the crystallinity of graphene fibers [62, 63]. Nevertheless, the approach always needs a fairly long duration at high temperatures (>12 h) (>2000 °C), which leads to reduced productivity and increased manufacturing costs. In addition, the conformational order of graphene sheets in fibers is an important parameter of fiber properties. Studies have shown that higher plate orientation can reduce structural disorder and improve the compactness of graphene fibers, resulting in better electrical and mechanical performances [64–66].

Joule heating is an advanced approach for efficiently manufacturing thermostable or metastable materials. During joule heating, current flows through the conductor and reaches about 3000 K in a few seconds, overcoming the dilemma of traditional heating [67, 68]. More recently, a couple of researches have shown that electric-induced joule heating can significantly restore conjugated frameworks in defective graphene and significantly increase the conductivity and carrier mobility of graphene film in a very brief time [15, 69–71]. Fine control of microstructure and wafer alignment can be achieved by adjusting the current applied to the sample. However, the stationary electrode configuration and limited sample size with a centimeter scale are incompatible with mass manufacturing [71–73]. Therefore, the continuous preparation of independent highly conductive graphene film in a fast and scalable manner remains challenging. The researchers are also keen to develop an effective strategy for synthesizing graphene films and improving the conformational order of the inner graphene sheets.

1. Roll-to-roll preparation of graphene fiber

In 2021, Yi Cheng et al. applied joule heating approach to the manufacture of graphene fiber and innovatively integrated dynamic design to achieve successive production. Joule heating can transform defective GO fibers into highly crystalline graphene fibers with extremely short processing time (20 min at ≈2000 °C) and low energy consumption (≈2000 kJ m^{-1}). In addition, a current-induced electric field was used for the first time to orient graphene sheets and increase their conformational order. The increase of the conformation order of the sheet can further improve the properties of the fiber. Compared to traditional thermoannealed graphene fibers (TGFs), joule thermal graphene fibers (JGF) exhibit higher orientation coefficient, excellent electrical conductivity, and tensile strength. This scalable approach paves the innovative way for the rapid and successive production of highly efficient graphene fibers, which will promote practical applications in power cables, electromagnetic shielding, and wearable electronics. Besides, the current-induced effect in this process can be applied to the conformational manipulation of component units in the assembled materials, which can further improve their macroscopic properties [74].

The homemade dynamic joule heating (DJH) facility for successive manufacture of graphene fibers includes four main components, containing a joule heating module, roll-to-roll equipment, vacuum device, and gas supply system, as shown in Fig. 5.13a. During the DJH procedure, a bundle of RGO fibers is coiled on two winding rollers and passed through the joule heating zone at programmed speed. At the same time, infrared cameras were used to monitor the temperature of JGFs. When the input current flows through the graphene fiber, the temperature of the fiber between the two electrode pulleys can rise to a stable value in a few seconds because of the joule heating effect. JGFs manufactured by joule exhibit uniform gray contrast and high flexibility when heated, as shown in Fig. 5.13b. High temperature in DJH procedure can effectively remove carbon impurities and improve the crystallinity of the fiber. The impurity removal was characterized by X-ray photoelectron spectroscopy (XPS), which showed that the peak value of O1s at ≈533 eV decreased gradually with the increase of temperature until it finally disappeared at 2000 °C, which was consistent with the corresponding EDS and thermogravimetric analysis results. Compared with RGO fiber (RGOFs), JGF treated at 2000 °C showed negligible D-peak in Raman spectra and uniform 2D peak intensity mapping, manifesting that the prepared JGFs were of high quality.

In summary, this innovative DJH method enables ultra-fast continuous preparation of graphene fibers. High temperature treatment times of graphene fibers in joule heating can be an order of magnitude shorter than conventional heat treatment times previously reported. During joule heating, the current flowing through the JGFs can manipulate the conformational order of the graphene sheet units, forcing them to align along the fiber axis, leading to a significant increase in the Hermann orientation factor, conductivity, and TGFs tensile strength. The DJH approach with current-induced

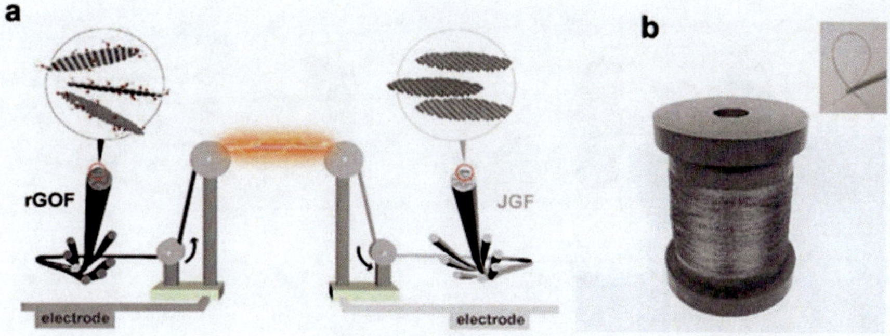

Fig. 5.13 Graphene fibers prepared by DJH **a** Schematic diagram of DJH system. The RGOFs wound on the supply roll are successively moved to the receiving roll through the joule heating zone; **b** 100 wire JGF rolls, inline diagrams showing the flexibility of prefabricated JGFs

effect can be extended to the successive synthesis of general assembly materials, the conformation control of component units can be realized synchronously, and its macroscopic properties can be designed.

2. Roll to roll to prepare carbon film

In 2019, Yingjun Liu et al. reported a successive, high-throughput, and energy-efficient strategy for annealing graphene films in a roll-to-roll manner with intense joule heat. The electrical input is applied directly to the sample via a rotating graphite roller. Since most of the applied powers are converted to thermal energy of the film, its energy utilization rate should be much higher than that of traditional electric furnaces. The resulting films have outstanding thermal conductivity (1285 ± 20 W/mK) and conductivity (4.2×10^5 S/m), both comparable to graphene films annealed in conventional furnaces at the same temperature [75].

In order to achieve successive reduction of graphene films by joule heating, Liu et al. carefully designed a roll-to-roll preparation system, including three parts of film supply, current, and film collection, as shown in Fig. 5.14a. The chemically RGO film is successively passed through rotating graphite rollers driven by two controllable micromotors. A programmed voltage is applied to parallel graphite rollers by two brushes, causing direct current (DC) to flow through the RGO film, resulting in temperature rise because of joule heating effect, as shown in Fig. 5.14b. By adjusting the rotation rate of the roller, the film can adapt to the size reduction caused by joule heating. The graphite roller not only acts as a rotating electrode but also applies mechanical pressure on the joule-heated RGO film to obtain pure graphene film with solid structure.

Figure 5.14c shows that the intensity of light emitted from the film increases as the input power increases, manifesting a gradual increase in temperature due to joule heating. During programmed heating, emission spectra of 350 ~ 950 nm were collected by a fiber optic spectrometer, as shown in Fig. 5.14d. By fitting the spectrum to the gray body radiation equation, the temperature of the film was extracted, and the temperature of the sample was recorded continuously and in real

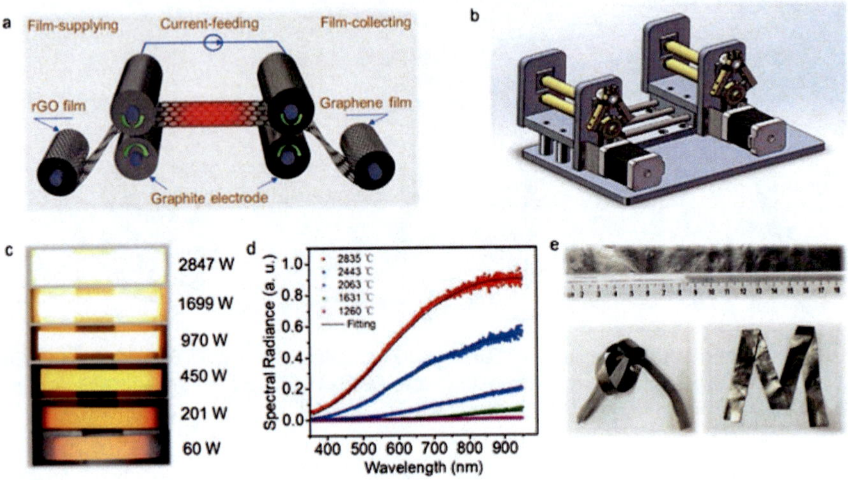

Fig. 5.14 **a** Schematic diagram of graphene film production by joule heating pressure roll; **b** schematic diagram of successive electric heating facility driven by two fast control motors; **c** thin-film snapshots heated by increasing electrical energy from the bottom up; **d** spectral radiation measurements of thin films with different power inputs; **e** photographs of joule-heated graphene film, showing good flexibility and foldability

time by two non-contact two-color hyper thermometers to ensure the accuracy of measurement. The sublinear curve shows that heating efficiency decreases at higher temperatures because more heat is consumed by heat conduction and radiation at higher temperatures. Joule heating film temperature up to 2835 °C, considering the safety and stability of roll-to-roll facility, the actual maximum temperature in joule heating process is set at 2443 °C.

The production cost of thermal conductivity film primarily focuses on the power consumption during the carbonization/graphitization procedure. Traditional electric furnace thermal annealing can achieve high thermal conductivity but require longer time (at least two days) and enormous energy costs because of slow heating and cooling speeds. The average power per hour of the micro high temperature graphitization furnace is about 50 ~ 70 kW/h, and the overall heating time is around 6 ~ 10 h. In contrast, the roll-to-roll joule heating reduction approach is highly efficient, which can be completed in 10 min, and the total power consumption is less than 3 kW. It is clear that high joule heating in roll-to-roll mode is more time and energy efficient and cost-effective, particular for the production of experimental prototypes. Therefore, this flexible rolling method has great potential to replace the traditional electric furnace heating approach. This simple processing strategy allows graphene films to be fabricated on a large scale and flexibly, with potential applications in thermal management, energy batteries, and wearable electronics.

5.3 New Devices Based on HTS

5.3.1 Heater

With advances in micro/nanoscience and engineering, materials, structures, and facilities have been designed and developed at the micro/nanoscale. In order to study physical and chemical properties, particularly heat-related behaviors, it is of importance to offer high-temperature heating elements for target objects in small spaces. The shape and size of the heater can also be entirely dependent on the heating requirements of material synthesis and processing to maximize thermal efficiency while avoiding damage to components around the system. Therefore, it is necessary to design a minimal three-dimensional heating element to make the local distribution more uniform by precisely controlling the temperature. Microscale heating elements have been extensively studied and elaborated by utilizing microhot plates as heat sources [76–78]. A minimum heating element of 100 μm has been developed using photolithography. However, because of the poor high-temperature stability of its material, the minimized heater is restricted to a maximum temperature of 1000 K. Besides, these planar microhot plates are two-dimensional in shape and can only generate temperature differences in the target object along the direction of heating.

In 2016, Yonggang Yao et al. reported a printable heater that uses 3D printing of viscous graphene oxide (GO) aqueous solution to form a battery of complicated GO structures, which are then heated in joule via electric current to generate high temperatures up to 3000 K in a controlled manner. Rapid response to temperature rise, heating rate up to 20,000 K/s. The high viscosity of concentrated GO inks and the high-resolution of direct 3D printing allow control of size and shape, enabling the heater to be printed in any shape and size. The three-dimensional heater also shows good stability at high temperatures and can perform a stable temperature switch over 2000 cycles between indoor temperature and 2000 K. The outstanding performance and potential cost effectiveness of expandable 3D-printable manufactured heaters proved in this work can offer promising heating options for a series of applications, particularly at high temperatures and 3D shapes [79].

In this work, highly viscous GO dispersions in water were prepared using an improved Hammer method. Such GO dispersions with appropriate viscosity can be directly printed into different shapes by stratification controlled by a computer program, as shown in Fig. 5.15a. A three-dimensional horseshoe structure was printed in this study. The embedded image exhibits the tiny heater with a cavity diameter of 1.5 mm, much smaller than a quarter coin. The photo image of the horseshoe 3D heater is exhibited in Fig. 5.15b.

The 3D-printed horseshoe heater was thermally reduced at 873 K to achieve a suitable conductivity of 3 S/cm. RGO is then further reduced at high temperatures (up to 3000 K) using a simple and efficient joule heating procedure. The connections between RGO sheets mainly determine the resistance, which may result in local temperatures higher than adjacent regions. Local high temperatures at the junction can also effectively reduce resistance there. As described below, the special high

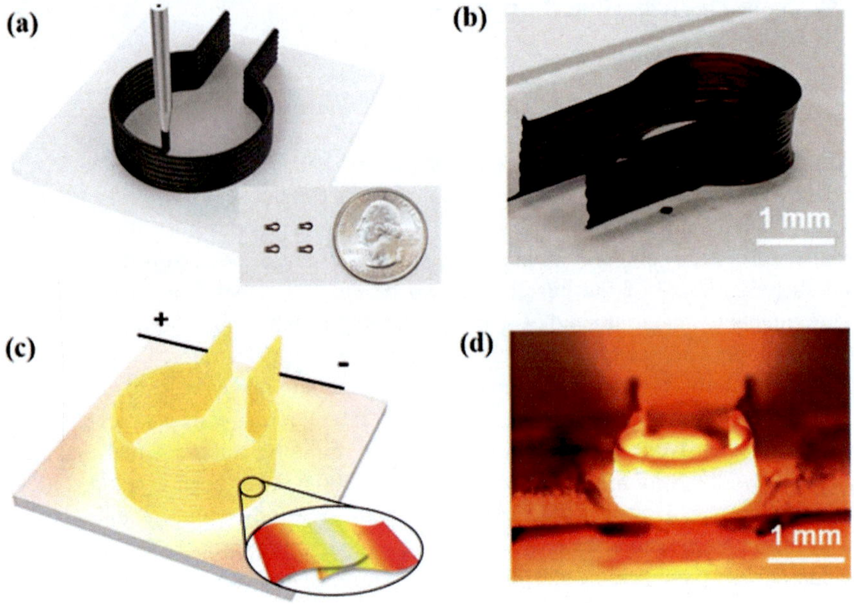

Fig. 5.15 Schematic illustration of 3D printable heater **a** 3D printing of RGO heater uses high concentration GO ink. The illustration shows an array of 3D-printed heaters with a space size of 1.5 mm; **b** photographic image of a printed horseshoe 3D heater; **c** the RGO heater is operated at high temperatures through a drive; **d** photographic images of 3D-printed RGO heaters operated at high temperatures

temperature reduction of the three-dimensional heater including RGO flakes effectively improves the conductivity. Figure 5.15c, d shows the schematic diagram and the physical drawing of high temperature heating, respectively. Due to the greater contact resistance, the contact area between the two RGO flakes is expected to experience higher temperatures and can be easily controlled by the applied current, inferred from the blackbody radiation. The unique characteristics of the heater come from the excellent printability of GO and its high temperature ability to reduce GO after high temperature reduction.

The annealing process of the heater is described in detail below. The 3D-printed heater was first freeze-dried to remove aqueous solvents and then pre-annealed in argon at 600 °C. The heater is then connected to the copper electrode for high-temperature reduction by joule heat, as shown in Fig. 5.16a. Silver paste is utilized to connect the RGO heater to the copper electrode, which is suspended above the ceramic substrate to avoid heat loss because of direct contact. Using the optical fiber on the RGO heater to collect the thermal radiation spectrum, the temperature of the heater based on the blackbody radiation is obtained. Figure 5.16b shows three-dimensional RGO heaters operating at different temperatures (1300 K, 1800 K, 2000 K) at different input power levels (0.5 W, 3 W, 6 W). As the input power increases, the heater heats up and lights up, and the higher the temperature, the brighter the light.

During the annealing process, the resistance continued to decrease, indicating that the high temperature joule heating further reduced RGO, so the resistance of the heater decreased 56 times, and the conductivity of the heater changed from 3 S/cm to 144 S/cm before and after the high temperature reduction. Compared with other reduction approaches, joule heating improves the reduction degree of GO by high temperature, high efficiency, and no chemical method.

XRD showed that the (002) peak became sharper after joule-heated high temperature reduction, manifesting a higher crystal structure because of the removal of defective atoms and the reconfiguration of carbon during high temperature reduction. After annealing, the sheet becomes strong and durable because of deformation caused by high temperature reduction. The SEM image of cross section also shows the porous structure of the RGO heater interconnecting with the RGO sheet. This is because 3D printing ink aqueous solvent after freeze drying dehydration, the 3D structure is already porous, and the thermal reduction of GO releases gas to further increase the porosity of the RGO heater. Poor contact between the porous structure and the RGO sheet reduces the conductivity of the RGO heater. Nevertheless, as mentioned earlier, these additional resistors generate joule heat for more efficient heating.

The RGO heater on a ceramic substrate can work effectively as a three-dimensional high temperature heater, as shown in Fig. 5.17a. Local temperatures at different locations of the 3D heater were measured by type K thermocouples at different input power levels, ranging from indoor temperature to 1500 K. Temperature distribution Fig. 5.17b shows the highest temperature in the center of the horseshoe heater, much higher than the surrounding area, manifesting that the designed RGO heater can locally generate and accumulate joule heat.

Three metals with different melting temperatures and shapes were placed in the RGO heater to observe their morphological changes under heating conditions, as shown in Fig. 5.17c. Tin spheres (T_m 505 K), aluminum columns (T_m 933 K), and copper columns (T_m 1357 K) achieved effective melting, and these materials are more spheroidized than the morphology changes before and after heating. This is because the RGO heater has a three-dimensional shape, which favors a more uniform temperature field for the metal column and ball, while the planar heating source has a huge thermal gradient along the heating direction and cannot achieve spheroidization.

Fig. 5.16 **a** Schematic diagram of high temperature annealing of 3D-printed RGO nanostructures and in situ temperature monitoring using optical fibers; **b** photographs of 3D-printed RGO heaters heated by joules at different powers

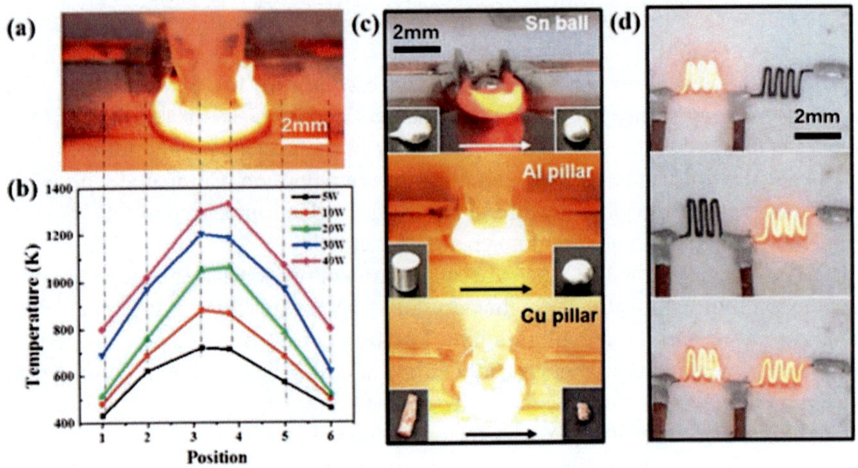

Fig. 5.17 3D-printed RGO heating element **a** 3D RGO heater with cavity diameter of 4 mm and **b** local temperature distribution at different power levels; **c** metal melting experiments of tin, aluminum, and copper at different power levels. The inlaid figure is demonstrated with printed RGO heaters for selective and controlled heating before and after morphological changes; **d** demonstration of controlled heating using a printed RGO heater

Due to the flexible design of 3D printing technology, the printed RGO heater can be integrated into the circuit as a controllable heating element, as shown in Fig. 5.17d, which is triggered by the applied DC voltage.

5.3.2 *Heating Probe*

Traditional pattern techniques such as atomic force microscope lithography are not suitable at high temperatures and are largely restricted by procedures and expensive equipment. Current local heating techniques, such as thermal scanning probe photolithography [80–82], despite their high spatial resolution, depend heavily on materials and equipment. In addition, the maximum sustainable heating temperature range (700 ~ 1000 °C) is restricted, and the processing rate is usually low. Hence, the development of a high precision high temperature heat source with superior heating and pattering capabilities is an ideal choice for thermally driven micro/nanofabrication, which is also a prominent challenge.

In 2021, Qiangliang Zhi et al. reported a 3D-printed RGO thermal probe triggered by electric joule heating as a high-temperature heat source with obviously improved temporal and spatial resolution for accurate thermal pattern drawing and synthesis of nanomaterials [83]. In this work, the extrusion-based direct ink-writing (DIW) approach is used for probe fabrication because of its simple procedure and great compatibility with various nanomaterials. In terms of materials, RGO was chosen

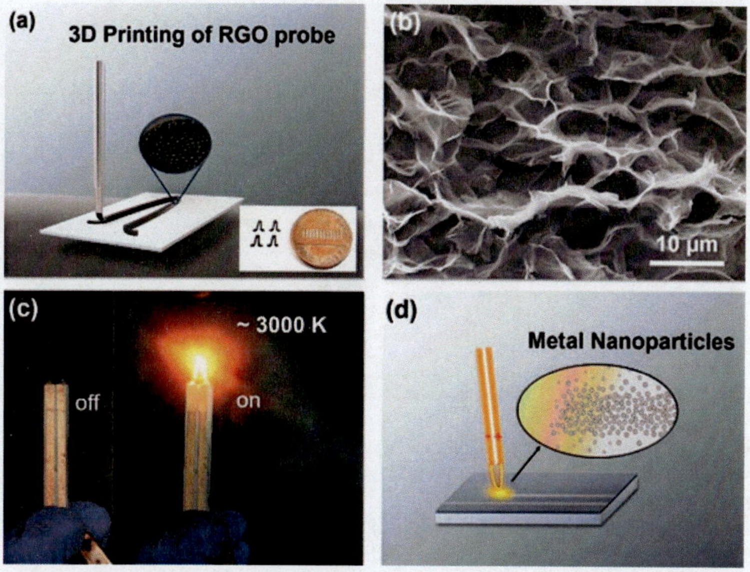

Fig. 5.18 3D printing of RGO probe **a** Schematic diagram of 3D printing of RGO probe. Embedded image: microstructure of RGO probe (middle) and array of printed RGO probe (bottom); **b** SEM image of cross section of RGO probe; **c** optical image of RGO thermal probe operating in "off" and "on" states (≈3000 K); **d** schematic diagram of thermal patterned metal nanoparticles on the substrate by fast radiative heating with a high temperature RGO thermal probe

because of its superior flexibility and thermal stability in 3D printing [84]. As shown in Fig. 5.18a, a water-based RGO ink is used to print an RGO probe with adjustable dimensions and geometry. The honeycomb-like microstructure shown in Fig. 5.18b achieves relatively high contact resistance compared to the dense structure, thus achieving effective joule heating. Besides, the strongly bonded honeycomb structure has great mechanical integrity and improved maneuverability [85, 86].

The printed RGO probe (after thermal reduction) is assembled in an alumina ceramic handle for electric joule heating. Figure 5.18c shows the RGO thermal probe driven by 500 mA current. The RGO heating probe has excellent thermal stability, reaching high temperatures up to ≈3000 K with a heating/cooling speed of $\approx 10^5$ K s^{-1} and high time resolution in the millisecond order. This thermal probe, with its microscale tip characteristics and outstanding heating capability, provides an effective strategy for accurate thermal profiling and nanomaterial synthesis. As a proof, metal nanoparticles, containing platinum (Pt) and silver (Ag), were fast synthesized on a carbon nanocarbon substrate by RGO thermal probes using a flexible "direct writing" procedure without the need for masks or extra development steps, as shown in Fig. 5.18d. The high spatial (submillimeter) and temporal (millisecond) resolution and excellent stability make the RGO heating probe an unbalanced heating source, ideal for local heat treatment, and thermally driven micro/nanomanufacturing.

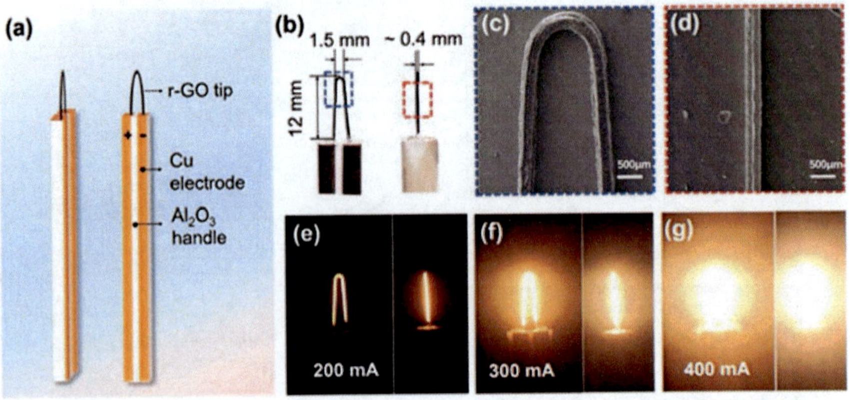

Fig. 5.19 High temperature RGO thermal detector triggered by electric joule heating **a** side (left) and front (right) views of the RGO thermal probe attached to an alumina ceramic handle attached to a copper electrode; **b** optical images of front (left) and side (right) views of the RGO thermal probe; **c, d** SEM images of front view and side view of RGO probe; **e–g** front and side views of the RGO thermal probe at 200, 300, and 400 mA, respectively

The following details the device. To achieve position control during electric joule heating, the printed RGO probe is fixed to the alumina ceramic handle and connected to the copper electrode by silver paste, as exhibited in Fig. 5.19a. This configuration can be used to manually heat a specific location or in combination with automation to achieve programmatically controlled local heating. Optical images show that the thickness, tip width, and height of the probe are about 0.4, 1.5, and 12 mm, respectively, as shown in Fig. 5.19b. SEM images further describe the small size of the probe, whose width is ≈400 μm, as shown in Fig. 5.19c, d. Figure 5.19e–g shows the front view of the RGO thermal probe, with front and side views of different drive currents of 200, 300, and 400 mA applied, respectively. As the input current increases, an increase in detector brightness indicates a rise in temperature.

In order to reveal the influence of high temperature joule heating on the structure and conductivity of RGO probe, the changes of RGO materials before and after joule heating were characterized. After stable electric joule heating at a driving current of 400 mA (≈2400 K), the conductivity of the RGO probe is significantly increased because of the reduction and graphitization of RGO at high temperature. Structurally, compared to the printed RGO probe after joule heating, the RGO probe shows a high strength graphite (I_G) band and a low strength disorder (I_D) band. This manifests a high degree of graphite after joule heating. The I_D/I_G ratio decreased from 0.95 for the printed RGO probe to 0.3 for the joule heat-reduced RGO probe, manifesting a significant reduction in the number of defects. In addition, XRD showed that the joule-heated RGO probe showed a peak at ≈26.316°, while the printed RGO probe showed a relatively wide peak at ≈23.158°, confirming the high graphite crystallinity caused by high joule heating.

As mentioned above, this high temperature RGO thermal probe has the following obvious merits over traditional heating technologies:

(1) Flexible heating probe shape design, with microscale tip characteristics;
(2) Highly controlled and stable heating with high time resolution (ms).

These properties make the RGO thermal probe an ideal heat source for the synthesis and manufacture of materials, such as patterns of metal nanoparticles (NPs), which require precise spatial and temporal control of the heating procedure.

As a proof of concept, thermal shock synthesis of Pt and Ag nanoparticles was performed on substrates loaded with metal salt precursors by rapid heating and quenching using an RGO probe device, Fig. 5.20a. To accurately control the position and movement of the RGO thermal probe, the alumina ceramic handle of the facility is fixed on a three-dimensional mobile platform, as shown in Fig. 5.20b. SEM shows that the area swept by the probe contains many Pt NPs evenly dispersed on the surface of the RGO substrate, with an average size of 36 nm. High-resolution TEM images further show the 0.19 nm lattice fringe spacing of NPs, corresponding to the (200) crystal planes of the face-centered cubic (FCC) platinum crystal structure. The synthesis of metal nanoparticles by fast thermal shock contains two processes: thermal decomposition of metal salts at high temperature and nucleation and growth of metal particles. In this procedure, the high temperature drives the decomposition of salt, and the brief duration of radiation heating is key to avert coarsening and aggregation. In contrast, traditional heating methods are difficult to have precise time and space control and tend to lead to the aggregation of nanoparticles. Since this is a fast and non-equilibrium synthesis procedure, metallic NPs often appear spherical to minimize surface energy. The size and quantity of NPs can be controlled by adjusting the duration of heating and the amount of metal salts loaded. In addition, because the temperature of the thermal probe is precisely controlled, a variety of metallic NPs can be synthesized by adjusting the temperature above the decomposition temperature of the corresponding salt precursor.

The great tunability of the RGO probe allows flexible control of the scanning path of the probe tip to accurately map the metal NPs in selected areas on the substrate. As a proof of concept, s-shaped curved NPs patterns were prepared on the surface of CNF films. The results show that it is feasible to rapidly integrate metallic NPs in selected areas using RGO probes in thermo scan probe photolithography, which makes it possible to rapidly construct plasma intensifiers, as shown in Fig. 5.21. The patterning capability and resolution of NP achieved by this method depend on the probe size, the distance between the probe tip and the substrate surface, and the probe temperature. Smaller probes naturally create finer patterns with higher resolution. Given the scope of this work, detailed optimizations, and fine-tuning of RGO probes will be carried out in future, as well as their combination with probe temperature, precursor load, and radiation distance at the minimum patterned size. In addition, this simple and rapid thermal patterning process also offers a potentially effective strategy for restoring the pattern of the GO film to the RGO electrode as shown in Fig. 5.21.

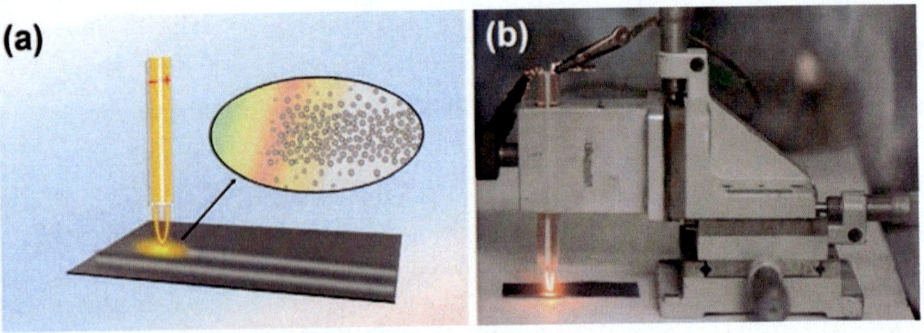

Fig. 5.20 Preparation of platinum and silver nanoparticles with RGO thermal probe **a** schematic diagram of displaying thermal patterns of metal nanoparticles on the surface of carbon nanoplates through a fast radiative heating procedure with high temperature RGO thermal probe; **b** image and operation of a device for manufacturing platinum NPs on RGO films utilizing a high temperature thermal probe triggered by electric joule heating

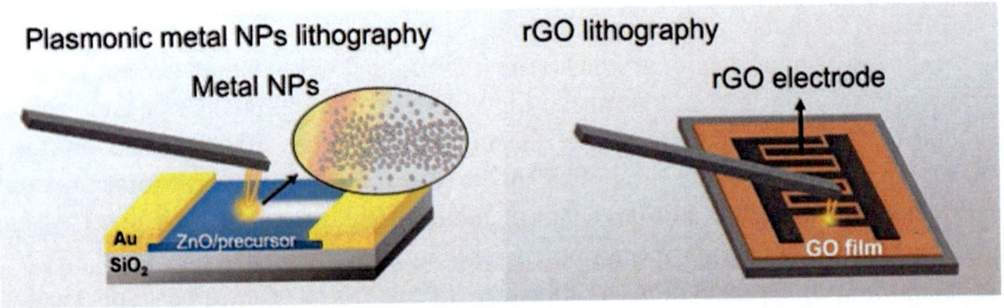

Fig. 5.21 Schematic diagram of preparing plasma FET based on RGO probe thermal scanning lithography and selective reduction of GO to interlaced RGO electrodes

The combination of 3D printing and electric joule heating provides an effective strategy for high precision thermal driven micro/nanofabrication with precise temporal and spatial resolution. Depending on heating requirements, the size and geometry of the RGO probe can be easily adjusted by changing print parameters, providing great flexibility for heating operations and potentially expanding its applications.

5.3.3 *Lighting Equipment*

Modern lighting technologies based on fluorescent lamps and light-emitting diodes (LED) are optimized for high luminous efficiency, but there are other challenges [87]. For instance, fluor fluorescent lamps include mercury vapor and must be disposed of as hazardous waste, while typical LED have multi-layer structures that require relatively complex and expensive manufacturing processes. Both techniques are limited to simple tubes and planes and do not include arbitrary and dynamic shapes such as flexible folding that need extreme mechanical strains on the constituent materials. On the other hand, tiny tungsten filament bulbs based on blackbody radiation are still ubiquitous because of their simplicity and low cost. Carbon materials have high melting and sublimation temperatures in vacuum, which are ideal for lighting in vacuum to reduce parasitic heat loss [88, 89], and have higher radiant power density due to their high emissivity. In fact, the utilization of carbon in a vacuum has long been used for lighting, dating back to Edison's first commercial light bulb using carbonized bamboo. Nevertheless, the previous carbon materials, whether large or nano, have failed to compete with the effective lighting of tungsten wires. Although a large number of carbon nanotube and graphene-based incandescent light sources have been demonstrated [90–92], merely, a few have reported operating temperatures above 2000 K, and none has achieved the 3000 K required to compete with tungsten filament lamps [93, 94].

In 2016, Wenzhong Bao et al. reported a carbon nanopaper (thickness 0.5 ~ 1 μm) composed of monolayer graphene oxide (GO) nanosheets and 10 wt.% single-walled carbon nanotubes. A particular two-step reduction procedure transforms the paper into a highly graphitized, highly conductive form, resulting in eye-catching lighting properties. The first thermal reduction step mixed moderately reduced graphene oxide (RGO) carbon paper with carbon nanotubes, while the second higher heating step further purified RGO-CNT paper and improved the crystallinity of the graphene plane. After two-step reduction, the RGO-CNT carbon paper showed outstanding graphene-like Raman spectra with a high-density structure and a record-high conductance at indoor temperature (2400 S/cm). This highly purified RGO-CNT paper can be maintained at a temperature of about 3000 K, which is significantly higher than paper made from carbon nanotubes or graphene alone. The ultra-thin and free-standing RGO-CNT paper is as efficient in vacuum as, and possibly better than, tungsten wires in argon. More importantly, the flexible, lightweight, and ultra-thin properties of the RGO-CNT belt make the material a prime candidate for flexible electronics, wearable devices, and powerful radiant heaters with fast high temperature response [95].

The RGO-CNT paper is suitable for various low-cost fabrications technologies and flexible shape requirements that are not available in traditional lighting technologies (tungsten, LED, or fluorescent). The different shapes and formats can be made by an additive method (i.e., 3D printing) or a subtraction method (i.e., paper trimming and cutting), then folded and bent, as exhibited in Fig. 5.22a. Figure 5.22b shows the programmed printing of a carbon nanofilament for illumination. The outstanding mechanical strength and flexibility of nanocarbon paper also allow cutting and pruning into different shapes. Nanocarbon paper is cut into a circular shape many times, which can be stably lit. It is also very flexible and lightweight, 10 times lighter than a tungsten filament that produces the same amount of light. In addition, it can be significantly bent even during illumination, as shown in Fig. 5.22c. Thin nanocarbon paper can also be losslessly folded and effectively spot bright while maintaining shape, as shown in Fig. 5.22d, e. These characteristics are not possible with other lighting facilities such as luminescent lamps, fluorescent lamps, and tungsten filament.

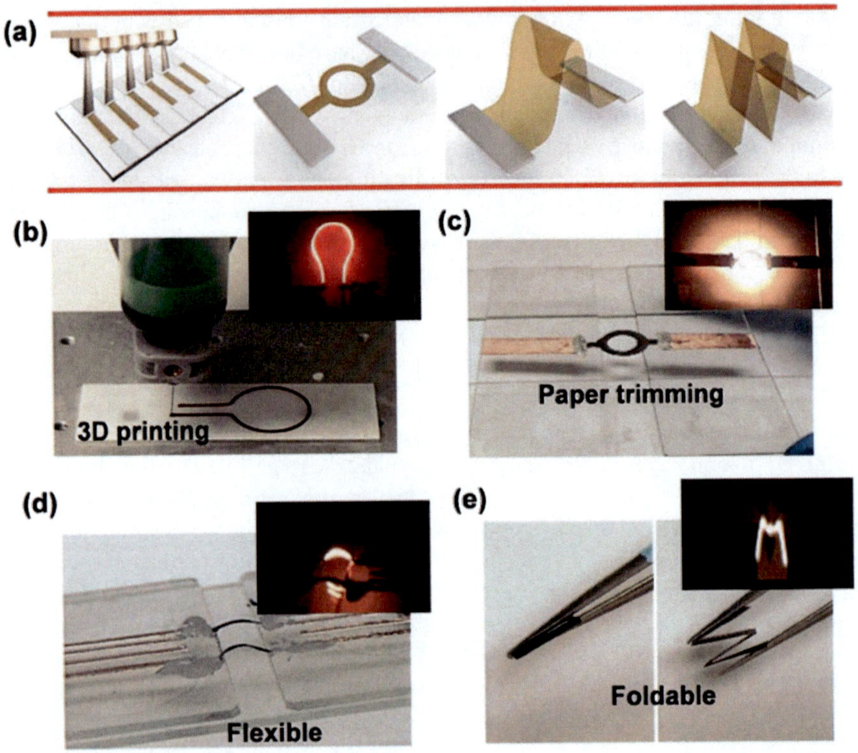

Fig. 5.22 **a** Schematic diagram of carbon nanoparticles in various forms; **b** 3D-printed carbon nanopaper for light bulbs; **c** nanocarbon paper printed by Muller rod can be pruned into different shapes; **d** flexible carbon nanopaper for lighting; **e** foldable carbon nanopaper for lighting

This ultra-pure and highly conductive RGO-CNT nanocarbon paper is manufactured from solution printing followed by a particular self-stabilized annealing procedure. Thin paper exhibits better thermal radiation properties, containing visible light emission, than any other carbon nanomaterial. Although the luminosity in the visible range is lower than that of LED or fluorescent lighting, it has the advantage of a broad spectrum and a variety of shapes and forms through simple and low-cost ink printing manufacturing. Under optimized processing conditions, RGO-CNT carbon paper exhibits visible lighting efficiencies close to or even higher than tungsten filament bulbs, providing a new solution for future unconventional lighting/heating needs. In addition, it can also be used for heater of material rapid heat treatment, thermionic emission cathode, heater of windshield deicing, etc.

5.3.4 Propellant

Nanoparticles attached to conductive substrates are ubiquitous in electrochemical energy storage, catalysis and high-energy facilities, and conductive substrates offer fast electron transport for device operation. Typically, nanoparticles are first synthesized and then assembled into a conductive carbon matrix [96]. For reactive nanoparticles, obvious oxidation often occurs during these two steps, which raises significant challenges for the effective use of these nanoparticles [97].

In 2016, Yanan Chen et al. reported an ultra-fast (2 ms) process for the preparation of uniformly distributed nanoparticles in a conductive RGO matrix by directly heating a metal/semiconductor-RGO film to a high temperature ($\approx$1700 K or higher), as shown in Fig. 5.23a. Microsized metal or semiconductor particles used as precursors are melted under high-temperature joule heating and then self-assemble into nanoparticles as they cool down. The aggregation of nanoparticles driven by surface energy minimization is inhibited by defects in the RGO layer in and out of the plane, as shown in Fig. 5.23b. These defects, as barriers to atomic migration, ensure the uniform distribution of nanoparticles. Because the RGO during heating is impermeable to O_2 and H_2O, the particles in the matrix are protected from the oxidation of RGO tablets. The defect site and high melting temperature (stable to 3300 K) of the RGO sheet can be utilized as a perfect substrate for this high-temperature process. In addition, this high-temperature method is suitable for in situ preparation of nanoparticles without condensation and surface oxidation and is suitable for any material with melting point below 3300 K. In this research, Chen et al. demonstrated the rapid synthesis of aluminum (Al), silicon (Si), tin (Sn), gold (Au), and palladium (Pd) nanoparticles. Nanomaterial forms of these materials have been widely utilized in energy storage, optics, sensing, and catalysis [98]. This section focuses on the application of nanoparticles prepared by HTS method in high-energy materials.

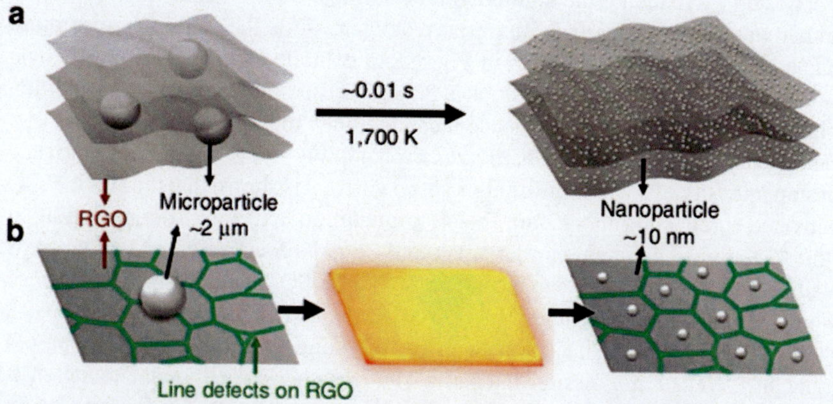

Fig. 5.23 Schematic diagram of in situ nanoparticle self-assembly procedure **a** Direct joule heating drive of self-assembly of particles in conductive RGO network matrix (left) into B10ms (right); **b** mechanism of nanoparticle formation

The nAl-RGO was chosen here to prove its superior performance as a high-energy material. Aluminium nanoparticles were prepared by the above method with a synthesis process of about 10 ms, which is much faster than many traditional synthesis approaches such as ball milling, physical vapor deposition, and liquid phase chemical synthesis. The maximum temperature of the nAl-RGO sample was 1730 K during ultra-fast heating. After cooling, the average size of the obtained aluminum nanoparticles is 10 nm.

These well-dispersed nanoparticles without surface oxidation in conductive substrates are particularly suitable for utilization as high-energy materials in several potential applications. High-energy materials are materials that can convert stored chemical potential energy into heat energy very quickly (e.g., propellants and explosives). There was a recent attempt to use nanoparticles in such applications, as they have been shown to burn faster and at lower ignition temperatures than microsized nanoparticles. Nevertheless, reactive nanoparticles are also likely to agglomerate and rapidly lose surface area because of aggregation and sintering upon heating, which reduces the advantages of utilizing nanoparticles (i.e., increased specific surface area and shorter diffusion distance) [99]. The natural oxide layer is another problem, as even the typical 2.5 nm layer formed in contact with air can occupy a large portion ($\approx$30%) of the mass of nanoparticles [100]. Besides, one of the primary practical issues for nanomaterials is safety, especially in averting unplanned ignition events. Therefore, an ideal energy system would stabilize the nanostructure to be insensitive to accidental ignition and prevent aggregation and oxidation, thus realizing the full potential of the nanoscale.

The energy properties of the material are shown in Fig. 5.24. The first image shows a high-speed video frame of activated 1:1 nAl-RGO, which reacts vigorously in an oxygen-rich atmosphere. The sample (1700 mm long, 410 mm wide, 4 μm thick) is connected to a wire with a silver paste and heated by an electric current (70 mA). Prior to the reaction, the sample was held at 1500 K, 0.001 atm for 2 min and activated by reducing the alumina on the surface of RGO. After activation, the chamber was permeated with 3 atm of pure oxygen, while the material temperature remained high. The second image in Fig. 5.24a exhibited the reaction of a single RGO under the same conditions, which is far less drastic, indicating that the addition of aluminum nanoparticles to the system can facilitate the reaction.

The quantization degree of reactivity is shown in Fig. 5.24b. By plotting the time emission intensity of each frame of high-speed video, it is shown that the single RGO is activated (preheating in vacuum; black line) and no activation (direct heating at 3 atm; Green line) all show similar weak emission. Non-activated nAl-RGO films (blue line) show moderately enhanced emission on RGO, manifesting the existence of some active Al nanoparticles. In contrast, activated membrane materials (i.e., materials that have been heat treated prior to adding oxidants; red line) show a significant increase in emission intensity. The hypothesis for this phenomenon is that after the initial synthesis of the nanoparticles, the aluminum nanoparticles on the outer surface of the RGO are oxidized when the film is exposed to air, reducing reactivity when the sample is heated directly without activation. In the activation process, the oxide is reduced by high temperature and the existence of carbon, leaving

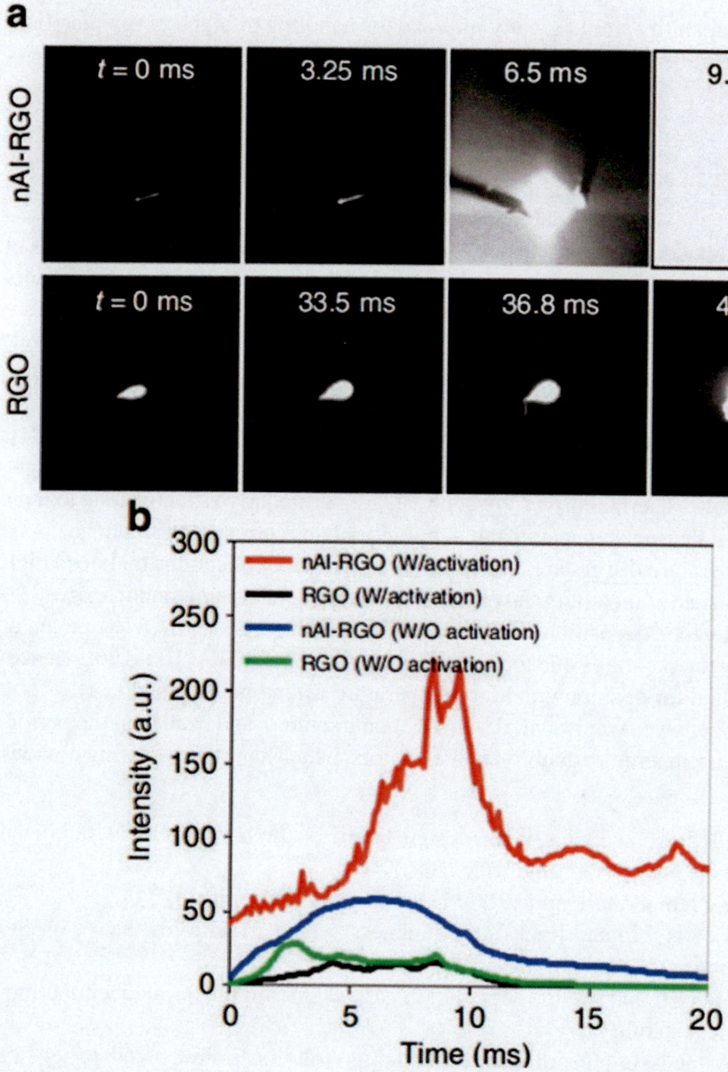

Fig. 5.24 Application of aluminum nanoparticles as high-energy materials in RGO networks **a** High-speed video frames, exposure time 142 ms, activated nAl-RGO and RGO, 3 atm O_2. The last frame shown is the last frame with peak composite intensity; **b** standardized composite intensity of video frames (20 ms exposure) for nAl-RGO and RGO samples

a highly reactive nanofuel that can completely take part in the reaction. Hence, RGO matrix offers a way to overcome the physical stability of nanoparticle oxidation and antiaggregation.

In terms of safety, the difference between the two composite cases suggests that the material's energy release can be adjusted in the field in a manner similar to a switch. This capability would greatly improve the handling of high-energy materials and enable the utilization of materials that are nominally too dangerous to use.

5.3.5 *Drive*

Actuator is a device that can change shape under different stimuli and is widely utilized in lots of fields, particularly robotics and sensors [101–103]. There are lots of stimulation approaches to achieve good driving effect, such as electrostatic, electrochemical, compressed gas, piezoelectric, light, heat, solvent, and steam. However, electroactive polymers (EAP) using electrostatic devices and piezoelectric actuators need higher voltages. Electrochemical actuators based on double capacitors require the use of electrolytes, which involves a complex sealing process [104]. Pneumatic actuators produced by compressed gas are restricted by complex high-pressure systems. Thermal drives, by contrast, are particularly attractive due to their material-based thermal expansion and simplicity of structure and mechanism.

High-temperature drivers are attractive for a series of applications that need high temperatures, such as metallurgical processing, single-crystal semiconductor growth, and chemical vapor deposition. Most conventional actuators can merely operate at relatively low temperatures due to material properties (<1000 K) [104–106]. Hence, a simple method for dealing with high-temperature actuators is needed.

In 2016, Chengwei Wang et al. designed, manufactured, and evaluated the performance of high-temperature double-layer actuators, which have the following obvious advantages:

(1) Thin and flexible double-layer design based on high-temperature stabilized materials (CNT and boron nitride (BN));
(2) Operating temperature up to 2000 K in inert gas environment;
(3) The device is thin and light (total thickness ≈11 μm) with low heat capacity and response time of 100 ms;
(4) Stable after 10,000 cycles, degradation can be ignored due to reduced thermo-mechanical stability;
(5) Low-cost and simple processing, including solution printing and paper-like cutting.

Proved high temperature actuators have the potential to be utilized as temperature sensors, mechanical switches, and robotic devices in a series of high temperature procedures [107].

In this work, two high-temperature stable materials (CNT and BN) were selected to create a double-layer actuator that can operate at temperatures up to 2000 K, too high a temperature will cause most materials to melt or evaporate. A significant increase in operating temperature enables the actuator to produce large deformation. Through selective joule heating, the U-shaped CNT film with a thickness of 10 μm is heated to 2000 K within 100 ms, at which point the actuator bends toward it, as shown in Fig. 5.25. When the current is removed, the actuator is rapidly cooled to

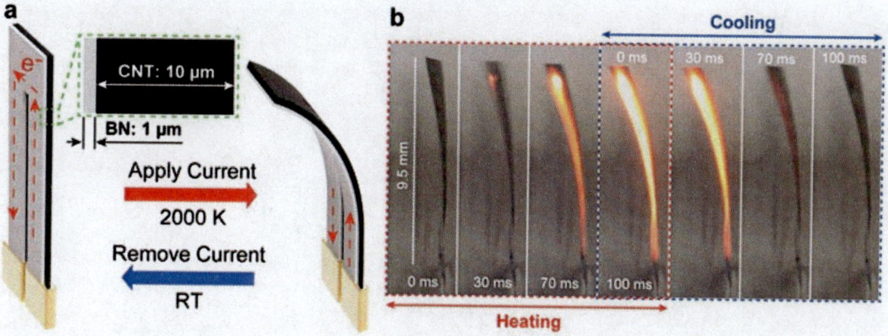

Fig. 5.25 CNT-BN double-layer high-temperature actuator **a** Schematic diagram of the double-layer actuator, in which the gray layer is BN ($\approx$1 μm), and the black layer is CNT ($\approx$10 μm). The current flows in a U-shape; **b** the time-lapse image of the response of the double-layer actuator to heat during heating and cooling phases

indoor temperature by radiation and heat conduction. The large surface area and high thermal conductivity of carbon nanotube films allow for rapid cooling. The time-lapse image in Fig. 5.25b shows that the actuator can be driven at high speed. Joule heating requires approximately 100 ms to power the actuator to its maximum temperature, and after removing the current, the driver requires approximately 100 ms to cool and restore shape.

The outstanding performance of high-temperature drives is due to the following reasons:

(1) The mechanical strength of 10 μm thick CNT film is large enough to be used as a support layer for 1 μm thick BN film;
(2) Double layer CNT-BN film has high flexibility, so that it can be cut into U-shape;
(3) The conductive properties of CNT and the insulating properties of BN lead to selective heating, resulting in a large effective difference in thermal expansion of the bilayer;
(4) BN nanosheet films have poor thermal conductivity in the direction of the passing film, so BN films will not be heated by the CNT layer, which might further increase the thermal expansion difference between the CNT layer and BN layer, making the driver respond faster to heating.

In order to carefully evaluate the performance of CNT-BN double-layer actuators at high temperatures, the following experiments were carried out. The two ends of the U-shaped actuator are attached to the edge of the glass slide and connected to the power supply with silver paint. By precisely adjusting the power, the temperature of the actuator is increased from about 1000 K to about 2000 K. The hot and cold state of the actuator recorded by a digital camera in this range. Then, the deviation degree of the free end of the driver was measured by superposing hot and cold images, as shown in Fig. 5.26. The results show that the migration temperature curves have good linearity, indicating that the mechanical and thermal properties of CNT and BN layers remain stable in the temperature range from indoor temperature to $\approx$2000 K.

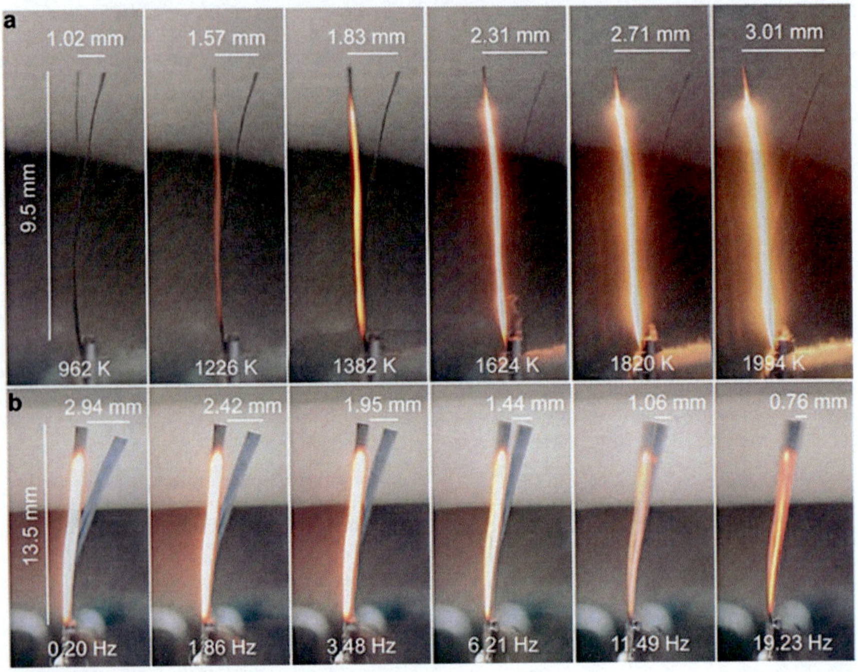

Fig. 5.26 Photos of CNT-BN double-layer actuators working at different temperatures and frequencies

The response rate of CNT-BN double-layer actuator is researched by frequency response measurement. Due to limited heat transfer, actuator deformation is reduced as the driving frequency increases from ≈1 to ≈20 Hz, especially at the end of the cooling step. Nevertheless, the actuator still responds to heat at high frequencies and has a high response speed that can meet the sensitivity requirements of practical applications.

In summary, this high temperature flexible actuator based on CNT-BN double film makes the actuator suitable for many potential applications, such as switches and sensors, with fast response speed, detectable large failures, and reliable performance. Double actuator in high temperature applications is particularly attractive, because most of the semiconductor and metal-based traditional devices will become dysfunctional or melt, and this kind of CNT-BN high-temperature perform appliance has a high melting point and excellent heat and mechanical performance, can keep for a long time to work, to study the thermal properties and mechanical properties of high temperature materials provide new opportunities.

5.3.6 Thermoelectric Conversion Devices

At present, concentrated solar energy and high temperature waste heat radiation power generation are primarily restricted to photovoltaic panels and mechanical heat engines. For waste heat and solar heat applications, the current temperature limit is below 1500 K. In order to efficiently convert heat into electricity, a high operating temperature (T) is required to ensure a high Carnot efficiency (thermal engines operating between 3000 and 300 K can produce 90% Carnot efficiency). At higher temperatures, however, mechanical thermals become more complicated. Therefore, for these and other applications, there is an urgent need to develop thermoelectric (TE) facilities based on mechanical flexibility and sufficient materials that can be integrated into large-scale fabrication. In addition, due to the increased mass and the inherent Carnot efficiency and power density, the high power factor and high operating temperature make the material research and development severely limited, which deserves further attention [108–110]. For TE applications with flexible polymers, the maximum operating temperature is more limited ($\leq$500 K). Therefore, it is highly desirable to develop new materials that can operate at high temperatures while keeping their flexibility and TE performance.

Carbon-based nanostructures are attractive to thermoelectric conversion materials because of their large surface-to-volume ratio, such as surface-modified graphene, and because carbon is a sufficient element on Earth. Nevertheless, graphene is not considered a material with great promise for thermoelectric applications because of its high thermal conductivity. In addition, graphene is difficult to produce on a large scale. Solution treated RGO has been investigated as a thermoelectric material due to its potentially low thermal conductivity and the ease of handling of large nanosheets. However, thermoelectric materials based on RGO generally show low conductivity and thermoelectric properties and are considered unsuitable for thermoelectric materials [111].

In 2018, Tian Li et al. demonstrated that graphene oxide thin films (HT-RGO) reduced at high temperature showed significant TE properties. The power factor of the reported RGO device increases by an order of magnitude, with its record-high reduction temperature (3300 K) increasing the conductivity to the level of so-called electronic crystals [112], with a power factor of 54.5 μW cm^{-1} K^{-2}. Reducing films have minimal oxygen content and a reduction bandwidth of about 1 meV, much lower than previously reported values [113, 114]. Besides effectively converting thermal radiation into electricity, HT-RGO also enables each layer to have a wide band of light absorption with a very high absorption coefficient [84]. This is fundamentally different from typical thermoelectric materials, which need greater thickness to ensure complete radiant heat absorption.

Li et al. designed an experimental device to study the TE performance working at high temperature through radiation heating, as shown in Fig. 5.27. A 1 mm wide and 15 mm long HT-RGO strip serve as the TE material, while another RGO strip is placed asymmetrically below as the radiation source. The temperature gradient along the RGO band then produces an uneven carrier distribution. The RGO strip, a few microns thick, is a flexible plate that can be easily integrated into existing high-temperature systems.

In practical applications, waste heat recovery materials and lightweight equipment are needed to improve work efficiency. Optical properties of materials such as transmittance, absorbance, and reflectivity are important for overall system efficiency because of their role in coupling external heat sources to TE materials. Therefore, the optical properties of RGO films at different annealing temperatures were studied. A 50 nm thick GO film was coated on a quartz substrate and reduced at 1000 K, 2000 K, and 3000 K, respectively. With the increase of reduction temperature, the transmittance decreases significantly. The transmittance of 3000 K RGO film is less than 0.01%, indicating that the absorption coefficient of graphene is as high as expected (2.3% per layer). The relatively flat absorption spectrum drops gradually from 0.8 in the ultra-violet to 0.6 in the near infrared. In addition, the reflectance of 3300 K RGO film was also measured, and the high absorbance and spectra confirmed that 3300 K RGO film is an effective broadband radiation absorber. Hence, unlike other less absorbent TE materials, HT-RGO functions as an integrated solar radiation converter that efficiently absorbs radiant heat and converts it into electrical energy using a smaller overall thickness.

In concept Fig. 5.28, concentrated solar energy takes place of the blackbody radiation source described in Fig. 5.27. The concentrated solar line incident on the device passes through an optical transparent cover (such as quartz), and the concentrated area is parallel to the width of the HT-RGO film suspended on the substrate. The color diagram shows the temperature change throughout the membrane. The top layer of the RGO film will absorb concentrated sunlight and generate a large temperature gradient through the right end of the P-type HT-RGO and the left end of the tungsten to output high TE voltage. The printed RGO films are reduced as thermoelectric materials at high temperature and have outstanding broadband light absorption properties. By optimizing the thickness of HT-RGO, the film can absorb concentrated solar energy (up to 3000 K) incident on the film surface.

In conclusion, this unique high-temperature treated RGO film can become a thermoelectric material with good properties. Compared to the previously reported carbon-based thermoelectric materials, HT-RGO shows outstanding power factor

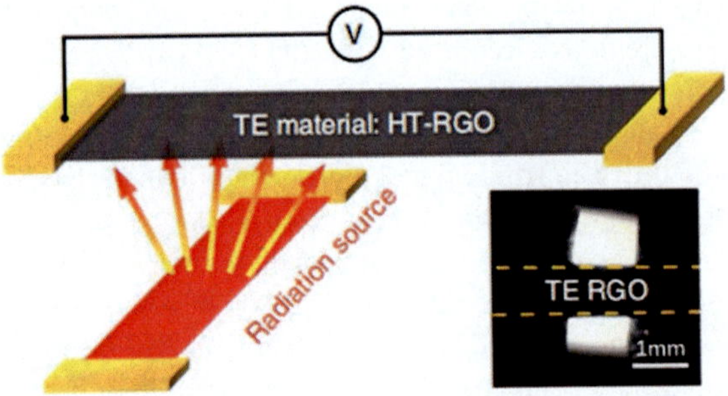

Fig. 5.27 Thermal radiation heating of HT-RGO film by the RGO band below it. The embedded image is the TE RGO band (dark) and the radiation band (bright) below

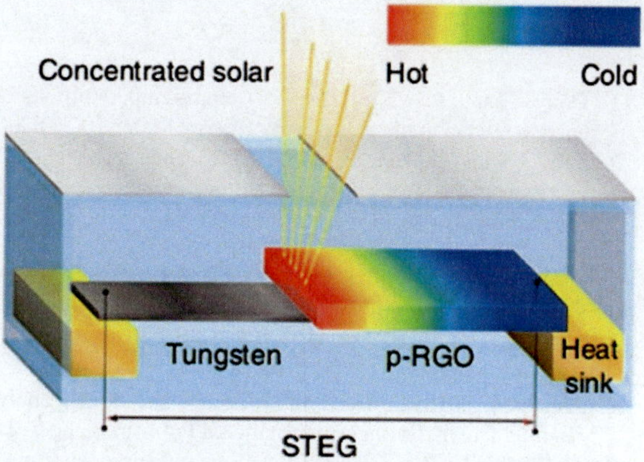

Fig. 5.28 Concept diagram of TE HT-RGO in a solar thermoelectric generator (STEG)

and can effectively absorb heat from 3000 K radiation sources in sub-micron thickness. It is not only flexible but also can provide reliable thermoelectric power output over a wide operating range (indoor temperature to 3000 K). Possible applications open up in lots of high-power energy systems, including radiant energy conversion, thermoelectric top cycling in power plants, and direct production from hydrocarbon combustion, in addition to concentrated solar energy. In addition, the conversion efficiency of the film can be further optimized by adjusting the doping level according to the required operating temperature.

5.4 Summary

Based on the characteristics of HTS, this chapter briefly introduces its application in new materials preparation equipment and new devices. In material preparation, HTS can combine the existing traditional methods to produce smaller size, better performance, more widely used nanomaterials, and low cost, fast process, controllable and have great potential in continuous production. In the application of new devices, based on the process advantages of HTS, practical applications such as three-dimensional microzone heating (heater) and precise heating (heating probe) can be realized. Based on the characteristics of HTS products, flexible lighting equipment, energy conversion materials (propellants), drivers, thermoelectric conversion materials, and a series of new devices can be developed.

To sum up, HTS has a promising future and will make great contributions in various fields, especially energy storage and conversion.

References

1. Dou S, Xu J, Cui X et al (2020) High-temperature shock enabled nanomanufacturing for energy-related applications. Adv Energy Mater 10(33):2001331
2. Leng J, Wang Z, Wang J et al (2019) Advances in nanostructures fabricated via spray pyrolysis and their applications in energy storage and conversion. Chem Soc Rev 48(11):3015–3072
3. Skrabalak SE, Suslick KS (2006) Porous carbon powders prepared by ultrasonic spray pyrolysis. J Am Chem Soc 128(39):12642–12643
4. Xu H, Guo J, Suslick KS (2012) Porous carbon spheres from energetic carbon precursors using ultrasonic spray pyrolysis 24(45):6028–6033
5. Xu Y, Liu Q, Zhu Y et al (2013) Uniform nano-Sn/C composite anodes for lithium-Ion batteries. Nano Lett 13(2):470–474
6. Zhu Y-M (2017) Software failure mode and effects analysis, failure-modes-based software reading. Cham, Springer International Publishing, pp 7–15
7. Jin Z, Xiao M, Bao Z et al (2012) A general approach to mesoporous metal oxide microspheres loaded with noble metal nanoparticles 51(26):6406–6410
8. Hong YJ, Son MY, Kang YC (2013) One-pot facile synthesis of double-shelled SnO_2 Yolk-shell-structured powders by continuous process as anode materials for Li-ion batteries 25(16):2279–2283
9. Liu J, Conry TE, Song X et al (2011) Nanoporous spherical $LiFePO_4$ for high performance cathodes. Energy Environ Sci 4(3):885–888
10. Xia B, Lenggoro IW, Okuyama K (2001) Correction: novel route to nanoparticle synthesis by salt-assisted aerosol decomposition. Adv Mater 13(23):1744–1744
11. Choi SH, Hong YJ, Kang YC (2013) Yolk–shelled cathode materials with extremely high electrochemical performances prepared by spray pyrolysis. Nanoscale 5(17):7867–7871
12. Jung DS, Hwang TH, Park SB et al (2013) Spray drying method for large-scale and high-performance silicon negative electrodes in Li-Ion batteries. Nano Lett 13(5):2092–2097
13. Motl NE, Mann AKP, Skrabalak SE (2013) Aerosol-assisted synthesis and assembly of nanoscale building blocks. J Mater Chem A 1(17):5193–5202
14. Chen P-C, Liu M, Du JS et al (2019) Interface and heterostructure design in polyelemental nanoparticles. Science 363(6430):959
15. Yao Y, Huang Z, Xie P et al (2018) Carbothermal shock synthesis of high-entropy-alloy nanoparticles. Science 359(6383):1489
16. Takahashi M, Koizumi H, Chun W-J et al (2017) Finely controlled multimetallic nanocluster catalysts for solvent-free aerobic oxidation of hydrocarbons. Sci Adv 3(7):e1700101
17. Yeh JW, Chen SK, Lin SJ et al (2004) Nanostructured high-entropy alloys with multiple principal elements: novel alloy design concepts and outcomes. Adv Eng Mater 6(5):299–303
18. Ye YF, Wang Q, Lu J et al (2016) High-entropy alloy: challenges and prospects. Mater Today 19(6):349–362
19. Buck MR, Bondi JF, Schaak RE (2012) A total-synthesis framework for the construction of high-order colloidal hybrid nanoparticles. Nat Chem 4(1):37–44
20. Nie Z, Li W, Seo M et al (2006) Janus and ternary particles generated by microfluidic synthesis: design, synthesis, and self-assembly. J Am Chem Soc 128(29):9408–9412
21. Ganguli AK, Ganguly A, Vaidya S (2010) Microemulsion-based synthesis of nanocrystalline materials. Chem Soc Rev 39(2):474–485
22. Li Z, Pradeep KG, Deng Y et al (2016) Metastable high-entropy dual-phase alloys overcome the strength–ductility trade-off. Nature 534(7606):227–230
23. Santodonato LJ, Zhang Y, Feygenson M et al (2015) Deviation from high-entropy configurations in the atomic distributions of a multi-principal-element alloy. Nat Commun 6(1):5964
24. Wang X, Huang Z, Yao Y et al (2020) Continuous 2000 K droplet-to-particle synthesis. Mater Today 35:106–114

25. Shatrova N, Yudin A, Levina V et al (2017) Elaboration, characterization and magnetic properties of cobalt nanoparticles synthesized by ultrasonic spray pyrolysis followed by hydrogen reduction. Mater Res Bull 86:80–87
26. Liang Y, Hou H, Yang Y et al (2017) Conductive one- and two-dimensional structures fabricated using oxidation-resistant Cu–Sn particles. ACS Appl Mater Interf 9(40):34587–34591
27. Wu C, Lee D, Zachariah MR (2010) Aerosol-based self-assembly of nanoparticles into solid or hollow mesospheres. Langmuir 26(6):4327–4330
28. Yang Y, Romano M, Feng G et al (2018) Growth of sub-5 nm metal nanoclusters in polymer melt aerosol droplets. Langmuir 34(2):585–594
29. Yang Y, Song B, Ke X et al (2020) Aerosol synthesis of high entropy alloy nanoparticles. Langmuir 36(8):1985–1992
30. Atwater HA, Polman A (2010) Plasmonics for improved photovoltaic devices. Nat Mater 9(3):205–213
31. Joo SH, Park JY, Tsung C-K et al (2009) Thermally stable Pt/mesoporous silica core–shell nanocatalysts for high-temperature reactions. Nat Mater 8(2):126–131
32. Liu L, Corma A (2018) Metal catalysts for heterogeneous catalysis: from single atoms to nanoclusters and nanoparticles. Chem Rev 118(10):4981–5079
33. Xu S, Chen Y, Li Y et al (2017) Universal, in situ transformation of bulky compounds into nanoscale catalysts by high-temperature pulse. Nano Lett 17(9):5817–5822
34. Salem A, Saion E, Al-Hada NM et al (2017) Simple synthesis of ZnSe nanoparticles by thermal treatment and their characterization. Results Phys 7:1175–1180
35. Zhang M, Dai Q, Zheng H et al (2018) Novel MOF-derived Co@N-C bifunctional catalysts for highly efficient Zn–Air batteries and water splitting. Adv Mater 30(10):1705431
36. Fei H, Dong J, Wan C et al (2018) Microwave-assisted rapid synthesis of graphene-supported single atomic metals. Adv Mater 30(35):1802146
37. Kumar J, Eraña H, López-Martínez E et al (2018) Detection of amyloid fibrils in Parkinson's disease using plasmonic chirality. Proc Natl Acad Sci 115(13):3225
38. Spielman-Sun E, Zaikova T, Dankovich T et al (2018) Effect of silver concentration and chemical transformations on release and antibacterial efficacy in silver-containing textiles. NanoImpact 11:51–57
39. Sun C, Li Y, Li Z et al (2018) Durable and washable antibacterial copper nanoparticles bridged by surface grafting polymer brushes on cotton and polymeric materials. J Nanomater 2018:6546193
40. Jiao M, Yao Y, Pastel G et al (2019) Fly-through synthesis of nanoparticles on textile and paper substrates. Nanoscale 11(13):6174–6181
41. Ding K, Cullen DA, Zhang L et al (2018) A general synthesis approach for supported bimetallic nanoparticles via surface inorganometallic chemistry. Science 362(6414):560
42. Li W, Liu J, Zhao D (2016) Mesoporous materials for energy conversion and storage devices. Nat Rev Mater 1(6):16023
43. Du Y-P, Héroguel F, Luterbacher JS (2018) Slowing the kinetics of alumina sol-gel chemistry for controlled catalyst overcoating and improved catalyst stability and selectivity. Small 14(34):1801733
44. Guo Z, Li J, Qi H et al (2019) A highly reversible long-life li–CO_2 battery with a RuP2-based catalytic cathode. Small 15(29):1803246
45. Jagadeesh RV, Murugesan K, Alshammari AS et al (2017) MOF-derived cobalt nanoparticles catalyze a general synthesis of amines. Science 358(6361):326
46. Liu Y, Qiao Y, Wei G et al (2018) Sodium storage mechanism of N, S co-doped nanoporous carbon: experimental design and theoretical evaluation. Energy Storage Mater 11:274–281
47. Wang S, Gao Y, Miao S et al (2017) Positioning the water oxidation reaction sites in plasmonic photocatalysts. J Am Chem Soc 139(34):11771–11778
48. Qiao Y, Yao Y, Liu Y et al (2020) Thermal shock synthesis of nanocatalyst by 3D-printed miniaturized reactors. Small 16(22):e2000509
49. Ma R, Zhang Z, Tong K et al (2017) Highly efficient electrocaloric cooling with electrostatic actuation. Science 357(6356):1130–1134

50. Majumdar A (2009) Helping chips to keep their cool. Nat Nanotechnol 4(4):214–215
51. Balandin AA (2011) Thermal properties of graphene and nanostructured carbon materials. Nat Mater 10(8):569–581
52. Balandin AA, Ghosh S, Bao W et al (2008) Superior thermal conductivity of single-layer graphene. Nano Lett 8(3):902–907
53. Seol JH, Jo I, Moore AL et al (2010) Two-dimensional phonon transport in supported graphene. Science 328(5975):213–216
54. Shahil KMF, Balandin AA (2012) Graphene-multilayer graphene nanocomposites as highly efficient thermal interface materials. Nano Lett 12(2):861–867
55. Peng L, Xu Z, Liu Z et al (2017) Ultrahigh thermal conductive yet super flexible graphene films. Adv Mater 29(27):1700589
56. Xin G, Yao T, Sun H et al (2015) Highly thermally conductive and mechanically strong graphene fibers. Science 349(6252):1083–1087
57. Lee C, Wei X, Kysar JW, Hone J et al (2008) Measurement of the elastic properties and intrinsic strength of monolayer graphene. Science 321(5887):1095–9203
58. Mayorov AS, Gorbachev R, Morozov SV, Britnell L et al (2011) Micrometer-scale ballistic transport in encapsulated graphene at room temperature. Nano Lett 11(6):1530–6992
59. Xu T, Zhang Z, Qu L (2020) Graphene-based fibers: recent advances in preparation and application. Adv Mater 32(5):1901979
60. Xu Z, Gao C (2011) Graphene chiral liquid crystals and macroscopic assembled fibres. Nat Commun 2(1):571
61. Xu Z, Sun H, Zhao X et al (2013) Ultrastrong fibers assembled from giant graphene oxide sheets. Adv Mater 25(2):188–193
62. Li Y, Zhu H, Zhu S et al (2015) Hybridizing wood cellulose and graphene oxide toward high-performance fibers. NPG Asia Mater 7(1):e150–e150
63. Zheng B, Gao W, Liu Y et al (2020) Twist-spinning assembly of robust ultralight graphene fibers with hierarchical structure and multi-functions. Carbon 158:157–162
64. Fang B, Chang D, Xu Z et al (2020) A review on graphene fibers: expectations, advances, and prospects. Adv Mater 32(5):1902664
65. Xin G, Zhu W, Deng Y et al (2019) Microfluidics-enabled orientation and microstructure control of macroscopic graphene fibres. Nat Nanotechnol 14(2):168–175
66. Xu Z, Liu Y, Zhao X et al (2016) Ultrastiff and strong graphene fibers via full-scale synergetic defect engineering. Adv Mater 28(30):6449–6456
67. Grosse KL, Bae M-H, Lian F et al (2011) Nanoscale Joule heating, Peltier cooling and current crowding at graphene–metal contacts. Nat Nanotechnol 6(5):287–290
68. Stanford MG, Bets KV, Luong DX et al (2020) Flash graphene morphologies. ACS Nano 14(10):13691–13699
69. Chen Y, Fu K, Zhu S et al (2016) Reduced graphene oxide films with ultrahigh conductivity as Li-Ion battery current collectors. Nano Lett 16(6):3616–3623
70. Chen Y-H, Lin K-H, Wang H-M et al (2018) The efficacies of esomeprazole- versus pantoprazole-based reverse hybrid therapy for Helicobacter pylori eradication. Advan Dig Med 5(1–2):26–32
71. Wang Y, Chen Y, Lacey SD et al (2018) Reduced graphene oxide film with record-high conductivity and mobility. Mater Today 21(2):186–192
72. Qu L, Liu Y, Baek JB et al (2010) Nitrogen-doped graphene as efficient metal-free electrocatalyst for oxygen reduction in fuel cells. ACS Nano 4(3):1321–1326
73. Yu D, Dai L (2010) Voltage-induced incandescent light emission from large-area graphene films. Appl Phys Lett 96(14):143107
74. Cheng Y, Cui G, Liu C et al (2021) Electric current aligning component units during graphene fiber joule heating. Advan Funct Mater 2103493
75. Liu Y, Li P, Wang F et al (2019) Rapid roll-to-roll production of graphene films using intensive Joule heating. Carbon 155:462–468
76. Hwang W-J, Shin K-S, Roh J-H et al (2011) Development of micro-heaters with optimized temperature compensation design for gas sensors. Sensors 11(3):2580–2591

77. Semancik S, Cavicchi RE, Wheeler MC et al (2001) Microhotplate platforms for chemical sensor research. Sens Actuators, B Chem 77(1):579–591
78. Son JM, Lee JH, Kim J et al (2015) Temperature distribution measurement of Au micro-heater in microfluidic channel using IR microscope. Int J Precis Eng Manuf 16(2):367–372
79. Yao Y, Fu KK, Yan C et al (2016) Three-dimensional printable high-temperature and high-rate heaters. ACS Nano 10(5):5272–5279
80. Garcia R, Knoll AW, Riedo E (2014) Advanced scanning probe lithography. Nat Nanotechnol 9(8):577–587
81. Gottlieb S, Lorenzoni M, Evangelio L et al (2017) Corrigendum: Thermal scanning probe lithography for the directed self-assembly of block copolymers (2017 nanotechnology 28 175301). Nanotechnology 28(28):289501
82. Ryu Cho YK, Rawlings CD, Wolf H et al (2017) Sub-10 nanometer feature size in silicon using thermal scanning probe lithography. ACS Nano 11(12):11890–11897
83. Liang Z, Yao Y, Jiang B et al (2021) 3D printed graphene-based 3000 K probe. Adv Funct Mater 31(34):2102994
84. Li T, Pickel AD, Yao Y et al (2018) Thermoelectric properties and performance of flexible reduced graphene oxide films up to 3000 K. Nat Energy 3(2):148–156
85. Peng M, Wen Z, Xie L et al (2019) 3D printing of ultralight biomimetic hierarchical graphene materials with exceptional stiffness and resilience. Adv Mater 31(35):1902930
86. Zhang Q, Zhang F, Xu X et al (2018) Three-dimensional printing hollow polymer template-mediated graphene lattices with tailorable architectures and multifunctional properties. ACS Nano 12(2):1096–1106
87. Reineke S, Lindner F, Schwartz G et al (2009) White organic light-emitting diodes with fluorescent tube efficiency. Nature 459(7244):234–238
88. Los JH, Zakharchenko KV, Katsnelson MI et al (2015) Melting temperature of graphene. Phys Rev B 91(4):045415
89. Savvatimskiy AI (2005) Measurements of the melting point of graphite and the properties of liquid carbon (a review for 1963–2003). Carbon 43(6):1115–1142
90. Armstrong MJ, O'dwyer C, Macklin WJ et al (2014) Evaluating the performance of nanostructured materials as lithium-ion battery electrodes. Nano Res 7(1):1–62
91. Liu Z, Bushmaker A, Aykol M et al (2011) Thermal emission spectra from individual suspended carbon nanotubes. ACS Nano 5(6):4634–4640
92. Singer SB, Mecklenburg M, White ER et al (2011) Polarized light emission from individual incandescent carbon nanotubes. Phys Rev B 83(23):233404
93. Dorgan VE, Behnam A, Conley HJ et al (2013) High-field electrical and thermal transport in suspended graphene. Nano Lett 13(10):4581–4586
94. Kim YD, Kim H, Cho Y et al (2015) Bright visible light emission from graphene. Nat Nanotechnol 10(8):676–681
95. Bao W, Pickel AD, Zhang Q et al (2016) Flexible, high temperature, planar lighting with large scale printable nanocarbon paper. Adv Mater 28(23):4684–4691
96. Jin Y, Zhang S, Zhu B et al (2015) Simultaneous purification and perforation of low-grade SI sources for lithium-Ion battery anode. Nano Lett 15(11):7742–7747
97. Su X, Wu Q, Li J et al (2014) Silicon-based nanomaterials for lithium-Ion batteries: a review. Adv Energy Mater 4(1):1300882
98. Chen Y, Egan GC, Wan J et al (2016) Ultra-fast self-assembly and stabilization of reactive nanoparticles in reduced graphene oxide films. Nat Commun 7:12332
99. Zong Y, Jacob RJ, Li S et al (2015) Size resolved high temperature oxidation kinetics of nano-sized titanium and zirconium particles. J Phys Chem A 119(24):6171–6178
100. Dreizin EL (2009) Metal-based reactive nanomaterials. Prog Energy Combust Sci 35(2):141–167
101. Li Q, Liu C, Lin Y-H et al (2015) Large-strain, multiform movements from designable electrothermal actuators based on large highly anisotropic carbon nanotube sheets. ACS Nano 9(1):409–418

102. Mosadegh B, Polygerinos P, Keplinger C et al (2014) Pneumatic networks for soft robotics that actuate rapidly. Adv Funct Mater 24(15):2163–2170
103. Zhang X, Yu Z, Wang C et al (2014) Photoactuators and motors based on carbon nanotubes with selective chirality distributions. Nat Commun 5(1):2983
104. Kim O, Shin TJ, Park MJ (2013) Fast low-voltage electroactive actuators using nanostructured polymer electrolytes. Nat Commun 4(1):2208
105. Haines CS, Lima MD, Li N et al (2014) Artificial muscles from fishing line and sewing thread. Science 343(6173):868
106. Zhang X, Pint CL, Lee MH et al (2011) Optically- and thermally-responsive programmable materials based on carbon nanotube-hydrogel polymer composites. Nano Lett 11(8):3239–3244
107. Wang C, Wang Y, Yao Y et al (2016) A solution-processed high-temperature, flexible, thin-film actuator. Adv Mater 28(39):8618–8624
108. Heremans JP, Dresselhaus MS, Bell LE et al (2013) When thermoelectrics reached the nanoscale. Nat Nanotechnol 8(7):471–473
109. Snyder GJ, Toberer ES (2008) Complex thermoelectric materials. Nat Mater 7(2):105–114
110. Zhao L-D, Dravid VP, Kanatzidis MG (2014) The panoscopic approach to high performance thermoelectrics. Energy Environ Sci 7(1):251–268
111. Xiao N, Dong X, Song L et al (2011) Enhanced thermopower of graphene films with oxygen plasma treatment. ACS Nano 5(4):2749–2755
112. Beekman M, Morelli DT, Nolas GS (2015) Better thermoelectrics through glass-like crystals. Nat Mater 14(12):1182–1185
113. Gao J, Liu C, Miao L et al (2016) Enhanced power factor in flexible reduced graphene oxide/nanowires hybrid films for thermoelectrics. RSC Adv 6(38):31580–31587
114. Wang W, Zhang Q, Li J et al (2015) An efficient thermoelectric material: preparation of reduced graphene oxide/polyaniline hybrid composites by cryogenic grinding. RSC Adv 5(12):8988–8995